# HOW TO MAKE SENSE OF STATISTICS

Sara Miller McCune founded SAGE Publishing in 1965 to support the dissemination of usable knowledge and educate a global community. SAGE publishes more than 1000 journals and over 800 new books each year, spanning a wide range of subject areas. Our growing selection of library products includes archives, data, case studies and video. SAGE remains majority owned by our founder and after her lifetime will become owned by a charitable trust that secures the company's continued independence.

Los Angeles | London | New Delhi | Singapore | Washington DC | Melbourne

# HOW TO MAKE SENSE OF STATISTICS

Stephen Gorard

$SAGE

Los Angeles | London | New Delhi
Singapore | Washington DC | Melbourne

# $SAGE

Los Angeles | London | New Delhi
Singapore | Washington DC | Melbourne

SAGE Publications Ltd
1 Oliver's Yard
55 City Road
London EC1Y 1SP

SAGE Publications Inc.
2455 Teller Road
Thousand Oaks, California 91320

SAGE Publications India Pvt Ltd
B 1/I 1 Mohan Cooperative Industrial Area
Mathura Road
New Delhi 110 044

SAGE Publications Asia-Pacific Pte Ltd
3 Church Street
#10-04 Samsung Hub
Singapore 049483

© Stephen Gorard 2021

Apart from any fair dealing for the purposes of research or private study, or criticism or review, as permitted under the Copyright, Designs and Patents Act, 1988, this publication may be reproduced, stored or transmitted in any form, or by any means, only with the prior permission in writing of the publishers, or in the case of reprographic reproduction, in accordance with the terms of licences issued by the Copyright Licensing Agency. Enquiries concerning reproduction outside those terms should be sent to the publishers.

Editor: Alysha Owen
Assistant editor: Charlotte Bush
Production editor: Martin Fox
Copyeditor: Richard Leigh
Proofreader: Neville Hankins
Marketing manager: Ben Griffin-Sherwood
Cover design: Sheila Tong
Typeset by: C&M Digitals (P) Ltd, Chennai, India
Printed in the UK

**Library of Congress Control Number: 2020941901**

**British Library Cataloguing in Publication data**

A catalogue record for this book is available from the British Library

ISBN 978-1-5264-1381-9
ISBN 978-1-5264-1382-6 (pbk)

At SAGE we take sustainability seriously. Most of our products are printed in the UK using responsibly sourced papers and boards. When we print overseas we ensure sustainable papers are used as measured by the PREPS grading system. We undertake an annual audit to monitor our sustainability.

This book is dedicated to the memory of Anthony John Gorard, a loving father and a kind and gentle man, who sadly passed away on 14th November 2020, aged 93.

# CONTENTS

*Detailed contents* ix
*List of figures* xvi
*List of tables* xvii
*About the author* xx
*Preface* xxi
*Online resources* xxiii

**PART I   INTRODUCTION** 1

1  Why we use numbers in research 3
2  What is a number? Issues of measurement 9

**PART II   BASIC ANALYSES** 17

3  Working with one variable 19
4  Working with tables of categorical variables 36
5  Examining differences between real numbers 51
6  Significance tests: How to conduct them and what they do not mean 63
7  Significance tests: Why we should not report them 77

**PART III   ADVANCED ISSUES FOR ANALYSIS** 89

8  The role of judgement in analysis 91
9  Research designs 105
10  Sampling and populations 114
11  What is randomness? 131
12  Handling missing data: The importance of what we do not know 136
13  Handling missing data: More complex issues 162
14  Errors in measurements 170

## PART IV   MODELLING WITH DATA — 179

15  Correlating two real numbers — 181
16  Predicting measurements using simple linear regression — 194
17  Predicting measurements using multiple linear regression — 201
18  Assumptions and limitations in regression — 219
19  Predicting outcomes using logistic regression — 226
20  Data reduction techniques — 245

## PART V   CONCLUSION — 255

21  Presenting data for your audience — 257

*Glossary* — 270
*References* — 278
*Index* — 287

# DETAILED CONTENTS

| | |
|---|---|
| *List of figures* | xvi |
| *List of tables* | xvii |
| *About the author* | xx |
| *Preface* | xxi |
| *Online resources* | xxiii |

| | |
|---|---|
| **PART I  INTRODUCTION** | **1** |
| **1  Why we use numbers in research** | **3** |
| Introduction | 3 |
| Why everyone needs to know something about the use of numbers | 3 |
| The format and structure of the book | 5 |
| *A note on software* | 6 |
| *The structure of the book* | 6 |
| Suggestions for further reading | 8 |
| | |
| **2  What is a number? Issues of measurement** | **9** |
| Summary | 9 |
| The kinds of numbers in social science | 9 |
| *Categorical values* | 9 |
| *Real numbers* | 10 |
| Further distinctions between types of numbers | 12 |
| *Ordinal values revisited* | 12 |
| *Ratio versus interval* | 14 |
| *Discrete versus continuous* | 15 |
| Conclusion | 15 |
| Notes on selected exercises | 16 |
| Further exercises to try | 16 |
| Suggestions for further reading | 16 |

## PART II  BASIC ANALYSES    17

### 3  Working with one variable    19
Summary    19
Data preparation and outliers    19
Summarising categories    22
Summarising real numbers    24
    *A measure of spread*    26
Common distributions    30
Conclusion    33
Notes on selected exercises    33
Further exercises to try    34
Suggestions for further reading    34

### 4  Working with tables of categorical variables    36
Summary    36
Working with tables    36
Adding a third category    40
Comparing more than two categorical variables    42
Hidden disadvantage: A real example    44
Summarising patterns in tables    47
    *Some of the problems*    47
    *A solution*    48
Conclusion    49
Notes on selected exercises    49
Further exercises to try    50
Suggestions for further reading    50

### 5  Examining differences between real numbers    51
Summary    51
Measuring differences between categories    51
Standardising the difference    54
    *Proportionate differences*    55
    *Effect sizes*    56
Assessing the impact of an intervention: A real example    58
Alternative versions of the common effect size    59
Conclusion    60
Notes on selected exercises    61
Further exercises to try    61
Suggestions for further reading    61

## 6 Significance tests: How to conduct them and what they do not mean 63
Summary 63
Conducting a significance test: Chi-squared 63
Conducting a different significance test: *T*-test 67
What does a *p*-value mean? 69
    *What a p-value does not mean* 70
Rewriting research results to avoid *p*-value results 72
Two further common distributions 72
Conclusion 74
Notes on selected exercises 75
Suggestions for further reading 76

## 7 Significance tests: Why we should not report them 77
Summary 77
Assumptions of significance tests, again 77
The inverse logic of tests 78
Significance tests and the population 81
Bayes' theorem 83
Power calculations 83
Confidence intervals 84
    *What is a confidence interval?* 84
    *An example of why CIs do not work* 85
Conclusion 87
Suggestions for further reading 88

## PART III  ADVANCED ISSUES FOR ANALYSIS 89

## 8 The role of judgement in analysis 91
Summary 91
Introduction 91
Judging the trustworthiness of a research finding 92
Judging the meaning of a finding 95
Generality 99
The warrant principle 99
Writing about values: A real example 101
Conclusion 102
Notes on selected exercises 103
Further exercises to try 103
Suggestions for further reading 103

## 9 Research designs — 105
- Summary — 105
- Research questions and designs — 105
- The elements of design — 106
- The independence of research designs — 110
- Conclusion — 111
- Notes on selected exercises — 112
- Further exercises to try — 113
- Suggestions for further reading — 113

## 10 Sampling and populations — 114
- Summary — 114
- Authority and generality — 114
- Populations and samples — 116
- Working with populations — 118
- The kinds of sample — 119
  - *Random selection* — 120
  - *Stratified sampling* — 121
  - *Clustered samples* — 122
  - *Non-probability samples* — 123
- How big should a sample be? — 124
  - *Deciding on your sample size* — 125
  - *Reasons for using a smaller sample* — 127
  - *How not to pick the size of a sample* — 127
- Conclusion — 128
- Notes on selected exercises — 128
- Further exercises to try — 130
- Suggestions for further reading — 130

## 11 What is randomness? — 131
- Summary — 131
- What is randomness? — 131
- Problems with randomness — 132
  - *Random sampling* — 133
- Cluster randomisation — 134
- Conclusion — 135
- Suggestion for further reading — 135

## 12 Handling missing data: The importance of what we do not know — 136
- Summary — 136
- Illustrating the problems of missing data — 136

| | |
|---|---|
| Working the example | 139 |
| Comparing two or more variables with missing data | 141 |
| Summary so far | 141 |
| Prevent missing data as far as possible | 143 |
| Report missing data accurately | 144 |
| Analysis of missing data | 145 |
| Missing data in an evaluation of Switch-on: A real example | 148 |
| Sensitivity analyses | 148 |
| How not to deal with missing data | 152 |
| Weighting the responses | 153 |
| Complete case analysis | 154 |
| Complete values analysis | 156 |
| Replacing with default values | 157 |
| Conclusion | 158 |
| Notes on selected exercises | 158 |
| Further exercises to try | 160 |
| Suggestions for further reading | 161 |

## 13 Handling missing data: More complex issues — 162
| | |
|---|---|
| Summary | 162 |
| Modelling simple imputation | 162 |
| Multiple imputation | 164 |
| The theoretical nature of missing data | 165 |
| Conclusion | 168 |
| Suggestions for further reading | 169 |

## 14 Errors in measurements — 170
| | |
|---|---|
| Summary | 170 |
| Errors in measurements | 170 |
| Error propagation | 172 |
| Taking repeated measurements for reliability | 174 |
| Conclusion | 176 |
| Notes on selected exercises | 176 |
| Suggestions for further reading | 177 |

## PART IV  MODELLING WITH DATA — 179

## 15 Correlating two real numbers — 181
| | |
|---|---|
| Summary | 181 |
| Introduction to correlations | 181 |
| A real example of a negative correlation | 185 |
| Multiple correlations | 187 |

| | |
|---|---|
| Assumptions and limitations of Pearson's *R* correlation analyses | 188 |
| A real example of a small correlation: The summer-born problem | 190 |
| Conclusion | 192 |
| Notes on selected exercises | 192 |
| Further exercises to try | 193 |
| Suggestions for further reading | 193 |

## 16 Predicting measurements using simple linear regression — 194
| | |
|---|---|
| Summary | 194 |
| Simple linear regression | 194 |
| Conclusion | 200 |
| Notes on selected exercises | 200 |
| Further exercises to try | 200 |
| Suggestions for further reading | 200 |

## 17 Predicting measurements using multiple linear regression — 201
| | |
|---|---|
| Summary | 201 |
| Introduction | 201 |
| Interactions | 204 |
| Creating the simplest model: A real-life example | 206 |
| *Dummy variables* | 207 |
| *Running a true multivariate regression* | 208 |
| *Forward entry of variables* | 209 |
| *Biographical and other models* | 211 |
| Conclusion | 216 |
| Notes on selected exercises | 217 |
| Further exercises to try | 217 |
| Suggestions for further reading | 218 |

## 18 Assumptions and limitations in regression — 219
| | |
|---|---|
| Summary | 219 |
| Introduction | 219 |
| Limitations of multiple linear regression | 220 |
| Assumptions of multiple linear regression | 221 |
| The independence of cases and hierarchical clusters | 222 |
| Conclusion | 225 |
| Suggestion for further reading | 225 |

## 19 Predicting outcomes using logistic regression — 226
| | |
|---|---|
| Summary | 226 |
| Logistic regression | 226 |
| A worked example | 227 |

| | |
|---|---:|
| Refining the model | 232 |
| Two-stage logistic regression | 237 |
| Changing the reference category | 240 |
| The possible determinants of entering higher education: A real example | 240 |
| Conclusion | 243 |
| Notes on selected exercises | 244 |
| Further exercises to try | 244 |
| Suggestions for further reading | 244 |

**20 Data reduction techniques** — 245

| | |
|---|---:|
| Summary | 245 |
| What is factor analysis? | 245 |
| *Assumptions underlying factor analysis* | 245 |
| A worked example of factor analysis | 246 |
| *Concluding warning* | 251 |
| Reliability | 251 |
| Conclusion | 253 |
| Suggestions for further reading | 253 |

**PART V  CONCLUSION** — **255**

**21 Presenting data for your audience** — 257

| | |
|---|---:|
| Summary | 257 |
| Introduction | 257 |
| Generic issues of presentation | 259 |
| Presentation of numbers | 260 |
| Presentation of tables and graphs | 261 |
| Real-life examples | 262 |
| *Comparing means over time* | 262 |
| *Regression analysis* | 266 |
| Conclusion | 268 |
| Suggestions for further reading | 268 |

| | |
|---|---:|
| *Glossary* | 270 |
| *References* | 278 |
| *Index* | 287 |

# LIST OF FIGURES

| | | |
|---|---|---|
| 3.1 | Bar chart of quality grades for 170 hospitals | 24 |
| 3.2 | Histogram of costs for 170 hospitals | 26 |
| 3.3 | Histogram of assignment scores | 29 |
| 3.4 | Histogram of uniform distribution | 31 |
| 3.5 | Histogram of normal distribution | 32 |
| 5.1 | Comparing incomes for residents in north and south | 54 |
| 5.2 | An effect size viewed in terms of two normal distributions | 57 |
| 6.1 | Histogram of chi-squared distribution | 73 |
| 6.2 | Histogram of $t$ distribution | 74 |
| 12.1 | A flow diagram to record missing cases | 145 |
| 15.1 | Graph comparing height and finger length (imaginary data) | 182 |
| 15.2 | Graph comparing two random variables | 184 |
| 15.3 | Graph comparing mean years at school living in poverty and mean school attainment, Key Stage 4, North-East England | 186 |
| 15.4 | Attainment in terms of Key Stage 2 points, by age-in-year, England | 190 |
| 15.5 | SEN reporting by age-in-year, England, 2015 KS4 cohort | 191 |
| 16.1 | The relationship between height and index finger length (imaginary data) | 195 |
| 16.2 | Line of best fit for data in Figure 16.1 | 196 |
| 19.1 | Classification plot for logistic regression model predicting employment | 233 |
| 20.1 | Amount of variance covered by each factor in a simple factor analysis | 247 |

# LIST OF TABLES

| | | |
|---|---|---|
| 3.1 | Frequency of quality grades for 170 hospitals | 22 |
| 3.2 | Assignment scores for 10 students | 26 |
| 3.3 | Assignment scores for 10 students, with absolute deviations from mean | 27 |
| 3.4 | Assignment scores for 10 students, with squared deviations from mean | 28 |
| 4.1 | Marginal totals for dataset of residence and voting intentions | 37 |
| 4.2 | Expected values for dataset of residence and voting intentions | 38 |
| 4.3 | Observed values for dataset of residence and voting intentions | 38 |
| 4.4 | Percentages for dataset of residence and voting intentions | 40 |
| 4.5 | Marginal totals for dataset of residence and voting intentions, with missing values | 40 |
| 4.6 | Incomplete expected values for dataset of residence and voting intentions, with missing values | 41 |
| 4.7 | Complete expected values for dataset of residence and voting intentions, with missing values | 41 |
| 4.8 | Observed values for dataset of residence and voting intentions, with missing data | 42 |
| 4.9 | Percentages for dataset of residence and voting intentions, with missing data | 44 |
| 4.10 | Percentage of each FSM category with other background characteristics | 45 |
| 4.11 | Percentage attaining each qualification threshold by FSM status | 46 |
| 4.12 | Standard two-by-two table | 48 |
| 5.1 | Mean earnings by area of residence | 53 |
| 5.2 | Estimated impact of Switch-on Reading Programme – gain score | 59 |
| 6.1 | Chi-squared test result, voting by area of residence | 66 |
| 6.2 | Result of $t$-test for independent samples – income by area of residence | 69 |

| | | |
|---|---|---|
| 8.1 | A 'sieve' to assist in the estimation of trustworthiness | 94 |
| 9.1 | Four common research designs | 106 |
| 12.1 | Estimating the income of an achieved sample | 140 |
| 12.2 | The results of a well-being experiment | 146 |
| 12.3 | Initial scores for dropouts in a well-being experiment | 146 |
| 12.4 | Percentage of two occupational groups not reporting their income in a survey | 147 |
| 12.5 | Pupils allocated to groups but with no gain score, and reason for omission | 148 |
| 12.6 | Difference between incomes in two towns, achieved sample | 151 |
| 12.7 | Difference between incomes in two towns, with a counterfactual case | 151 |
| 12.8 | Ethnic group by political flexibility, before weighting | 153 |
| 13.1 | Two complete cases and one case missing one value | 163 |
| 15.1 | Pearson correlation of height and length | 183 |
| 15.2 | Pearson correlations of height, length and age | 188 |
| 16.1 | Model summary predicting finger length from height | 198 |
| 16.2 | Coefficients for model predicting finger length from height | 198 |
| 17.1 | Model summary predicting finger length from height and age | 202 |
| 17.2 | Coefficients for model predicting finger length from height and age | 203 |
| 17.3 | Model summary predicting finger length from height and age, with interaction | 205 |
| 17.4 | Coefficients for model predicting finger length from height and age, with interaction | 205 |
| 17.5 | Model summary predicting reading score after intervention | 208 |
| 17.6 | Coefficients for model predicting reading score after intervention | 209 |
| 17.7 | Model summary predicting reading score after intervention, forward entry | 211 |
| 17.8 | Coefficients for model predicting reading score after intervention, forward entry | 211 |
| 17.9 | Model summary predicting reading score after intervention, biographical | 214 |
| 17.10 | Coefficients for model predicting reading score after intervention, biographical | 214 |
| 17.11 | Coefficients for most parsimonious model predicting reading score after intervention | 215 |

| | | |
|---|---|---|
| 19.1 | Base model from logistic regression predicting employment | 230 |
| 19.2 | Full model from simple logistic regression predicting employment | 231 |
| 19.3 | Coefficients for variables in a simple regression model predicting employment | 232 |
| 19.4 | Model from Block 1 of logistic regression predicting employment | 238 |
| 19.5 | Coefficients for Block 1 of logistic regression model predicting employment | 239 |
| 19.6 | Model from Block 2 of logistic regression predicting employment | 239 |
| 19.7 | Coefficients for Block 2 of logistic regression model predicting employment | 239 |
| 19.8 | Percentage predicted correctly for HE entry, step by step | 241 |
| 19.9 | Coefficients for variables in final step of model predicting HE entry | 242 |
| 20.1 | A four-factor model of career choice influences | 249 |
| 21.1 | Probability of getting the disease, having tested positive | 259 |
| 21.2 | Difference in the commitment and exploration scores per domain before ($T1$) and after ($T2$) the guidance ($n = 45$) | 263 |
| 21.3 | Difference in the commitment and exploration scores per domain before and after the guidance, simpler | 264 |
| 21.4 | Difference in the commitment and exploration scores per domain before and after the guidance, even simpler | 265 |
| 21.5 | Results of multiple regression analysis related to prediction of life satisfaction | 266 |
| 21.6 | Results of multiple regression analysis related to prediction of life satisfaction, simpler | 267 |

# ABOUT THE AUTHOR

**Stephen Gorard** is Professor of Education and Public Policy, and Director of the Evidence Centre for Education, at Durham University (https://www.dur.ac.uk/). He is a Fellow of the Academy of Social Sciences, member of the ESRC Commissioning Panel for the Research Methods Programme, the British Academy grants panel, the MRC Adolescence Mental Health Expert Review Panel, and Lead Editor for BERA's Review of Education. He is a member of the Cabinet Office Trials Advice Panel as part of the Prime Minister's Implementation Unit. His work concerns the robust evaluation of education as a lifelong process, focused on issues of equity, especially regarding school intakes. He is author of around 30 books and over 1,000 other publications. He is currently funded by Nesta to evaluate its EdTech Testbed, and by the British Academy to look at the impact of schooling in India and Pakistan.

# PREFACE

This book arose from nearly 25 years of teaching social science research methods, and writing methods resources. The book is based partly on the publicly funded research capacity-building work I have conducted in the UK, USA and elsewhere. And it is partly based on advice and help from many colleagues over the years.

It is also about the things I have learnt from the experience of conducting hundreds of research projects across many fields, publishing substantive pieces from each project, and trying to explain what these projects did to as wide an audience as possible. I am saying that I am not primarily a writer about research methods, but I do know how to conduct research. For this reason, my advice to newer researchers may tend to differ somewhat from 'professional' methods writers, many of whom make their living telling us how to do research, but are seemingly less keen on doing research themselves. Research is generally much easier to do than is portrayed by them.

The book can be navigated in different ways by different readers, as explained in Chapter 1. At heart the book is an introduction to research analyses with numbers, suitable for undergraduate students and wider readership. Adding the more advanced chapters and sections makes it appropriately challenging for master's, doctoral and early-career researchers. Through its general style, and the range of coverage in worked examples, it is relevant to social scientists in anthropology, business, criminal justice, economics, education, geography, health studies, linguistics, politics, psychology, social work, sociology and other fields.

Above all, what really drove me to write this new and very different book is a relatively recent shift in wider views on the role of significance testing and associated ideas in social science. Significance tests are useless for practical analytical purposes, widely misunderstood, and almost universally misused. As noted in the book, many or most important authorities, journals, funders and fields of research have now abandoned the idea of using significance tests. There has been a noticeable recent change in understanding, and I hope we are seeing the end of these strange artefacts in common usage.

However, I have now found that some lecturers, researchers and students are asking 'what shall we do instead?'. What is this 'new' statistics? This is not really the right question. If we are doing something wrong in our everyday lives and suddenly realise that it is wrong, then we would stop. Put absurdly, imagine you were hitting someone over the head for no reason. If they asked you to stop, then it would make no sense to ask them what you should do *instead* of hitting them. Stopping doing the wrong thing is improvement enough. So it is with significance tests and their linked procedures. Not using them ourselves and not accepting them from others are big steps towards improving the use of statistics in social science.

Hopefully this book can become a sort of halfway house. It does explain a bit about what significance tests are and how to conduct them, as well as explaining why we should not conduct them or take notice of them. This is to assist readers who come across these archaic approaches when reading prior research. But it is hoped that in the future, instead of teaching each new generation what significance tests are and then critiquing them, authors might write books that do not mention them at all. The end has to start somewhere. This would provide more space in methods resources for discussion of issues relating to research quality and rigour, which should be at the heart of any analysis.

This may concern some lecturers and module leaders. I recently heard from a psychology tutor who said that while they agreed that we should all move away from significance testing towards the use of effect sizes, they were professionally unable to do so. Their argument was that they taught a course accredited by the British Psychological Society, and the BPS demanded that the course included significance testing. Fortunately for progress, this is incorrect. Neither the BPS accreditation regulations nor the QAA Subject Benchmark Statement for Psychology mention significance testing or indeed any specific kind of analysis. The statements are all formulated in general terms. If this misunderstanding is widespread, I hope that it can be corrected quickly. Similar kinds of concerns based on tradition mixed with misunderstanding, and perhaps fear, may be hindering progress in other areas of social science as well, including perhaps economics.

So, the majority of this book does not describe significance tests. Rather it discusses more valid, sensible techniques and craft tips for analysing or modelling data from populations and opportunity samples as well as randomised ones. It looks in detail at the kind of analytical considerations that should matter even for those of us who used to use significance tests. These vital considerations seem to have been forgotten in the vacuous parade of 'it's less than 0.05, good, I can publish', or 'oh no, it's more than 0.05, what shall I do now?'. In summary, this new book begins to shows what to do *instead*.

# ONLINE RESOURCES

*How to Make Sense of Statistics* is supported by online resources to help you become confident doing statistical analysis. Find them at: https://study.sagepub.com/gorard.

Practise techniques discussed in the book on **real-world datasets** in SPSS and Excel formats, which are accompanied by **screenshots** of using SPSS to conduct analysis and example **syntax** and **output**.

Work through **additional exercises** to feel secure in your knowledge of different statistical techniques.

Find helpful **guides for your chosen software** with author Stephen Gorard's list of handy resources.

# Part I
# Introduction

# 1
# WHY WE USE NUMBERS IN RESEARCH

## INTRODUCTION

This is a book about using and understanding numbers in social science research. It differs from other resources you may come across, by presenting the process of using numbers as being simpler, and more powerfully liberating, than is usually portrayed. It encourages all academics, researchers and students to use and understand the use of numbers in research. This is not a book about the conduct of research in general, or the process of numeric data collection more specifically. It does not require entering a scary **paradigm** called **quantitative research**, or making any kind of **epistemological** commitment, whatever that is (see the glossary for explanations of terms set in bold). Everyone can use numbers in research easily, just as they do in their everyday life. The focus of the book is on the use of numbers that we have good access to, rather than on any hypothetical numbers that we cannot see, or which might not even exist. This is not a book about maths, and there are no complex sections of equations or algebra. The examples presented mostly require only elementary arithmetic to understand them. Instead, the use of numbers, as dealt with here, requires imagination, care, appropriate scepticism and lots of critical judgement. Using numbers in this way is rewarding, valuable for society and enjoyable for the researcher. There is nothing to be scared of.

## WHY EVERYONE NEEDS TO KNOW SOMETHING ABOUT THE USE OF NUMBERS

It is not really possible to live your life without using numbers in some way. You may use numbers when handling money, telling the time, planning your day, booking a train, concert ticket or holiday, checking your speed, battery life, fridge temperature, or the weather forecast, watching a game show or home improvement programme, weighing ingredients for cooking, weighing yourself, checking your activity tracker, playing board games or cards, buying shoes, or just calling a telephone number. The examples are endless. Numbers are one of the types of information that are key to understanding and control in our lives. Eschewing them would be absurd, and would make having a co-operative life in society almost impossible.

Everyone knows this. Many, perhaps most, people use the numbers that are important in their lives in an uncomplicated manner, without too much problem, and without obvious resistance. This is true even if the same people say that they are not good at, or cannot abide, maths. Numbers of the kind we use regularly are *not* maths. To take just one example, you can take your partner's temperature if they are ill without being or becoming a mathematician. Again, everyone knows this. And everyone, even a mathematician, may miss their bus, forget they had something in the oven, or not notice that they have been given the incorrect change in a shop. Mistakes such as these do not just occur with numbers, and are perhaps an inevitable part of the everyday. They do not mean that the person involved cannot or should not use numbers. Most of us routinely process a huge amount of numeric information successfully, often without even realising it.

However, when some people become social science researchers an odd thing can happen. Numbers can become a focus for divisive but needless arguments about their validity. Many practising social science researchers seem to reject the use of the same sort of numbers that they are happy with in their non-social-science lives. And then they try to defend that rejection on the basis of talk about paradigms, methods identities, and epistemologies (Ralston et al., 2016). It is often the researchers who are least confident in maths, or have most anxiety about statistics, who are most likely to defend their non-use of numbers in research through an appeal to the supposed illegitimacy of statistics.

Whatever the reasons, this is not a sustainable position. It is not possible to do real-life social research without encountering numbers in the same way as we all do when we are not researching. It is as crucial to be appropriately critical of the use of numbers in research as it is in everyday life. The sceptical approach, encouraged in this book, is completely contrary to just rejecting all use of numbers. In order to decide whether to trust some numeric information or not, to be able to discriminate between the trustworthy research and the rest, we need to understand quite a bit about how those numbers are used and how they behave.

On the other hand, some commentators and funders have suggested that there should be more statistical ('quantitative') studies in social science research, because this form of evidence is said to be intrinsically preferable and of higher quality than other forms. This is again completely the wrong way of looking at it. One reason to encourage a greater awareness of statistical techniques among all researchers is that so-called 'quantitative' work is currently often very poor, but it can have considerable real-life impact while being largely unchecked by a wider, more cautious, readership.

There are other reasons why all researchers should learn something about techniques for research involving numbers. All researchers need to read and use the research of others, because all new studies involve some consideration of prior work in that field. This is impossible to do unless researchers know something about the conduct of research with numbers. Otherwise they may just accept all numeric research as valid, which is a big mistake, or reject all research with numbers, which is prejudiced and an even bigger mistake. Or they may accept/reject results on the basis of ideology, or whether they are happy with what the research reports having found. This would be the biggest mistake of all. It is not a social science approach to research.

The supposed schism between research using numbers (so-called 'quantitative') and research not using numbers (so-called **qualitative**) is at fault for much of this. There is no need to use such divisive terms to describe your research (Gorard with Taylor, 2004). These terms are generally used either by people who do not do research, and so have not realised that once you start researching everything is or could be useful data. Or these terms are used as defences by people who want to avoid dealing with research of one kind or another, because they do not want to work at a reasonable scale (Chapter 10), or perhaps do not see the need for judgement in analysing results (Chapter 8). Everyone reads text in much the same way, and everyone checks their change in a shop in much the same way. Researchers should just report what they did, and what they discovered (Chapter 21).

So, this book suggests a better and more fruitful way forward.

## THE FORMAT AND STRUCTURE OF THE BOOK

The book is based on a wide range of simple worked examples and exercises that introduce the various topics, terms and techniques for using numbers, gradually, and in an order that will build up your knowledge and skills. Some of the ideas may seem very simple for some readers at the outset, but it is often useful to re-examine our basic understanding of the foundations of research. Some of the later ideas, such as logistic regression techniques, may seem difficult at first for some readers unused to working with numbers. However, whatever your prior experience, working through each chapter in turn should lead to a logical and ordered understanding, even though each substantive chapter is also intended to be readable on its own.

Each substantive chapter starts with a summary of its contents or purpose, and includes exercises for the reader, some simple worked examples, and at least one example of real-life social science research using the ideas in that chapter. The chapters end, where appropriate, with notes on the exercises for use by readers, or by lecturers using the text as the basis for a course, a further exercise usually based on datasets available on the accompanying website, and notes on a few suggestions for further reading. The further reading is based on resources that I have seen. There will be many other useful books, articles and websites.

## A note on software

The graphs and other outputs from the examples in this book are based on working with software in widespread use, such as Excel and SPSS. These programs can help you to do the calculations for your numeric analyses easily. Excel is more widely available for most users, while SPSS provides more help with a wider range of analyses such as factor analysis, logistic regression and other more advanced regression models. However, reading the book does not depend on using either of these programs, or any other available software such as R or Stata. Nor is the book intended to be a manual on how to use any of them. There is a list of useful and simple resources for setting up and using analytical software (particularly SPSS) at the end of Chapter 3.

Each worked example has a related box that contains the steps needed to produce the example outputs, including a summary of the SPSS **syntax** (a sequence of steps, like computer code, to produce an analysis). The advantage of syntax is that readers can repeat each analysis while changing the names of the variables to suit their own research. Readers who want to can simply skip these bits, just by skipping the inset boxes. For those who want to pursue it, the datasets described are available to play with on the website accompanying this book. Use this material in the book or ignore it as you wish (it is all boxed off, separate from the main flow of the text). Maybe ignore it the first time, and then go back and try out some of the worked examples.

## The structure of the book

A few of the topics are covered by a pair of related chapters. The first chapter of each pair contains all that you really need to know about how to conduct a specific form of analysis or preparation for analysis, presented as simply as possible. The second chapter in each pair has a slightly more difficult or technical section, generally explaining why what is in the first chapter is really all you need to know. If you are feeling tentative then you can skip these more technical chapters (Chapters 7, 10, 12, 14 and 18), and this will not affect your understanding of later chapters in the book at all. These slightly more technical chapters are for the more confident and curious readers, and also for lecturers and tutors who might want to know why the book does not contain some of the difficult stuff that so many other statistical texts do. This book emphasises throughout the kinds of true analyses with judgement that follow from the relatively simple stages of totalling, averaging or modelling your data. These true analyses requires care, creativity, logic, scepticism and dedication. But they are not overly technical.

Chapter 2 looks at what numbers are, and how to interpret them. It introduces a simple classification of two types of numbers – real and categorical.

Chapter 3 outlines the common techniques for describing and summarising one variable in a dataset at a time, in a way that is clear and simple. It introduces graphs, frequencies, percentages, modes, means, and both the absolute mean deviation and the standard deviation. Chapter 4 describes analyses with two categorical variables, including cross-tabulations and odds ratios. Chapter 5 introduces analyses with one categorical variable and one real number, including effect sizes based on the differences between means.

Chapter 6 describes how to conduct two common significance tests, and what their results mean and do not mean. Although the book as a whole suggests not conducting significance tests in your own work, this chapter should help you when reading the work of other researchers who still uses this archaic approach. Chapter 7 is a more technical chapter, explaining in more detail why significance tests do not help us and can be misleading, and are therefore not presented in the rest of the book. Shorn of significance tests and the like, it becomes much easier to use, write about and understand numbers, while nothing of any value is lost.

Chapter 8 shows how all analyses, including those with numbers, require judgement on the part of the researcher. And that finding results like those described in Chapter 6 and 7 is only the start of any analysis, not its destination. This chapter begins to describe how we can judge the trustworthiness of research findings, and then their generality and meaning.

Chapter 9 reminds readers of the importance of research design. The design of any study should stem from the research questions to be addressed, and should lead naturally to the kind of analyses that will be needed. Whether the design is appropriate for the research questions being addressed is a key judgement to make, as part of deciding how trustworthy any research finding is.

Chapter 10 introduces the ideas and terminology associated with populations and sampling – the cases we obtain our measurements from. The scale and quality of the cases used in research represent another factor to take into account when judging how trustworthy any research finding is. Chapter 11 is slightly more technical, looking at the philosophical idea of randomness, as used in random sampling, and some of the rarer and more complex methods of sampling, and why these are not usually needed.

Chapter 12 describes how to detect, report and handle missing data. How much missing data there is, of what kind, and how this is handled, is another key issue to consider when looking at how trustworthy a research finding is. Chapter 13 is more technical, explaining why the simple approaches to handling missing data in Chapter 12 are sufficient for most of us, most of the time. Chapter 14 is again slightly more technical. It expands on what a measurement is, the idea of measurement errors, and it discusses the issue of trying to measure latent variables such as attitudes through repeated questions. The quality of measurements used in research is an important factor to consider when judging how trustworthy any research finding is.

Chapter 15 looks at correlation analyses involving two real numbers, with a focus on Pearson's correlation coefficient ($R$). This leads into Chapter 16 on simple linear regression modelling, using one predictor variable to predict a real number outcome. Chapter 17 describes multiple linear regression, using several predictors at once, to predict a real number outcome. Chapter 18 is more technical and describes the assumptions underlying linear regression in more detail.

Chapter 19 introduces logistic regression modelling using multiple variables of any kind to predict a categorical outcome. Chapter 20 outlines how to understand and conduct both a simple factor analysis and a reliability analysis.

Chapter 21 completes the book by stressing the importance of simplification when presenting findings, so that the widest possible readership can understand them. It is not possible to judge the trustworthiness of research unless it is reported fully and comprehensibly.

It is hoped that this book will have a wide readership, and it certainly has useful advice both for beginners and for more experienced 'numerati'. Doctoral researchers and lecturers might use the whole book. And everyone should read the first and last chapters. But perhaps an introductory undergraduate course could focus at first on the substantive Chapters 2–5, 8, 10, 12, 15 and 16. This should provide more than enough material for an introduction. A master's methods course might add Chapters 6, 9, 17, 19 and 20.

Running an analysis to find your first real result, or to create your first regression model, is an exciting step. Enjoy working with numbers.

## Suggestions for further reading

One of the problems when identifying further reading in each chapter is that so much of the literature on numeric analysis has been 'captured' by the use of significance tests and the like. So, many of the suggestions throughout are for only part of a book or resource.

As the name suggests, this book is an easy introduction:
Donnelly, R. (2007) *The Complete Idiot's Guide to Statistics*. London: Alpha Books.

This paper, aimed in particular at experimental biologists, argues that all researchers must grasp basic statistics (although I do not recommend all of the techniques advocated):
Vaux, D. (2012) Know when your numbers are significant, *Nature*, 492, 180-181.

Further discussion of why everyone should use numbers in research, and how easy this is in practice:
Gorard, S. (2006a) *Using Everyday Numbers Effectively in Research*. London: Continuum.

A commentary on the push to have more 'quantitative' work in social science:
Platt, J. (2012) Making them count: How effective has official encouragement of quantitative methods been in British sociology? *Current Sociology*, 60(5), 690-704.

# 2
# WHAT IS A NUMBER? ISSUES OF MEASUREMENT

## SUMMARY

This chapter introduces two different ways in which numbers can be used to describe things in social science - known as categorical (ordinal and nominal) and real numbers (counts and measures). These distinctions are relevant to how such numbers are presented and analysed in later chapters of the book, and are all that you need to know for the rest of the book. The second part of the chapter summarises some other distinctions between numbers that are less relevant but that you might meet when reading other research or methods resources.

## THE KINDS OF NUMBERS IN SOCIAL SCIENCE

This chapter introduces you to the different kinds of numbers that are used throughout the rest of the book. There are many more different ways that numbers could be classified (as outlined later). But to keep things simple, this book focuses on only two kinds of numbers. You already know this distinction. Think of a speed in miles per hour (a real number) and the house numbers on a street (used as labels). In research, as in everyday life, numbers are used in different ways or, perhaps more accurately, different kinds of numbers are used in various contexts.

### Categorical values

Numbers can be used simply instead of names or titles (like house numbers on a street). Athletes may have entry numbers in a sprint race, or horses may be numbered to help identify them in a horse race. But such numbers may not represent a starting or finishing order in the race, or indeed anything meaningful at all. They are just a kind of name.

The same might occur in research when a researcher records the ethnic origin of a survey respondent using numbers rather than words for some reason of convenience. Perhaps South Asian respondents are recorded as being in group 1, and East Asian respondents as group 2. This does not suggest that the East Asian group (2) has twice as much of something as the South Asian (1), or is more than the other group in any way. The numbers and their relative sizes are completely arbitrary, and would be just as meaningful if the coding system were reversed (or if the letters A and B were used instead of the numbers 1 and 2). We could call this type of number **categorical** (identifying categories) and **nominal** (meaning 'naming', here naming with numbers).

Some numbers in life and research, however, imply also a simple ordered relationship between themselves and other numbers in the same context. For example, an oven might have a dial with numbers from 0 to 5 on it. This does not mean that turning the dial to 2 makes the oven twice as hot as when turning it to 1. But these numbers and their order are not arbitrary. Setting the dial to 2 should make the cooker hotter than setting it to 1. Similarly, a researcher may ask respondents in a survey about their parents' highest level of educational qualification. If the researcher then codes the responses so that having an A level or equivalent (an exam taken mostly by 18-year-olds in England) is recorded as 3, and having an undergraduate degree is recorded as 4, this does not mean that a degree is 33% better than an A level, or that three degrees is equal to four A levels. But a degree is generally accepted as a higher level of qualification than an A level. We could call this type of number 'categorical' and **ordinal** (having an order to the categories).

All categorical information, whether ordinal or not, relates to categories only. Examples of these kinds of numbers therefore cannot be subject to arithmetic operations. It makes no sense to add together two house numbers on a street. The sex of a doctor could be a nominal category, and we cannot subtract the maleness of one doctor from the femaleness of another to find their difference in terms of sex. This restriction applies even where the categories are expressed as ordinal numbers.

## Real numbers

Other numbers used in social science are more clearly counts, or measurements, of something (like a speed in miles per hour). The number of horses in a race, and the number of respondents of South Asian origin, are both examples of counts – counts of how many objects there are. If 20 survey respondents are South Asian and 4 are East Asian, then it is true to say that there are five times as many South Asian as East Asian respondents in that survey. Arithmetic operations can be conducted using the frequencies of categorical data in this way. We could find a difference by subtraction between the number of male and female employees in a factory, or find the number of houses with a street number greater than 100.

On the other hand, the speed of a car in miles per hour is a measurement. The height of a horse in metres is a measurement. A horse that was 2 metres high would be twice as tall as a horse 1 metre high. We can do arithmetic with both counts and measurements.

**Image 2.1** The numbering of racehorses

So, **real number** variables are those that it makes sense to do arithmetic with. A simple test of identification would be: does it make sense for me to add or subtract these numbers? The number of years an employee has been employed in a factory is a real number. To find the difference in experience between two employees we could subtract two numbers, and find how many years more one employee had been at the factory than the other. We can do this because the scale we use to measure time with has equal intervals all the way along. The difference between 19 years and 20 years is the same as that between 1 and 2 years, for example. A car travelling at 40 mph is going twice as fast as one travelling at 20 mph.

For the purposes of this book we will be dealing with two overarching kinds of numbers. One group (real numbers) includes both counts and measurements that it makes sense to do arithmetic with, and we will treat this group in the same way for most purposes. In the other group, (categorical) numbers can only be used as names for categories or classes of things. These numbers have no real meaning, although they may have an order.

The types of numbers we are working with has implications for how they are best analysed, as shown in Chapter 3 and beyond.

### Exercise 2.1

a    In a study of the prices of different kinds of houses, there are two variables – types of houses and house prices. What type is each variable?
b    In a study of the highest qualifications obtained by four ethnic groups, there are two variables – qualification and ethnic group. What type is each variable?
c    In a study of the relationship between parents' annual income and school test scores, there are two variables – test scores and income. What type is each variable?

### Exercise 2.2

Imagine that you want to measure three things in your new study. These are each respondent's self-esteem, age in years, and family income. Which of these do you think it will be easiest to measure and will yield the most valid results? Which of these do you think it will be hardest to measure and will yield the least valid results?

## FURTHER DISTINCTIONS BETWEEN TYPES OF NUMBERS

The distinction between real numbers (measures and counts) and categorical values (ordinal and nominal) is all you are likely to need to know, in order to understand how to analyse and present numeric information. Even the distinction between ordinal and nominal may be superfluous for most analyses. However, you will meet other names for types of numbers in other research, and perhaps even some different ways to handle them in other methods resources. So, this part of the chapter looks at the distinctions a bit further to help you understand these additional terms when they arise. A new researcher (or an impatient reader) may want to skip this part, and move on to the first examples of analyses in Chapter 3.

### Ordinal values revisited

An ordinal value will be on a scale that can be analogous to something observed, as long as the thing observed has an order to it, but it genuinely does not increase or decrease either along a continuum or in regular jumps. It is the **isomorphism** or direct equivalence between the numbers and the things being **observed** that gives us a true measurement (Berka, 1983; Gorard, 2003a). An ordinal scale should not be used, and would not then be isomorphic, simply because of ignorance about the way in which the thing of interest varies. It should not be just a sloppy interval measure.

An occupational class scale based on prestige is a good example of an ordinal value, whereas the grades achieved in an examination might not be, because the grades are probably imposed on a real number scale of the marks in the examination. The real number scale would be preferable, if available, and the grades are a much weaker version of that (in measurement terms). But the occupational class scale is not imposed on such a clear underlying measurement. It is commonly a matter of judgement, not of measurement, to allocate **cases** to occupational class categories.

Even so, for most analyses, the order in an ordinal variable makes little difference when compared to dealing with nominal variables beyond the obvious. In calculating the number of students achieving a certain combination of grades, such as A–C, a teacher is not going to add in the frequencies of students in grade D, for example. In the same way, with a nominal scale representing full- and part-time employment and no employment, an analyst can calculate the number of all employed people by adding the first two categories. They are not going to add in the unemployed category, except by mistake. Knowing the ordinal nature of the measurement here makes no practical difference. Whether a grade C is genuinely different from a grade D, and where the cut-off point should be if so, are difficult questions. But they are not ones that are related to the difference between nominal and ordinal variables.

It is also important to realise that an intrinsic order is available for just about any set of categories. Categories could be ranked by popularity, or alphabetically, or in terms of prestige, for example. The numbers for some TV channels in England such as 101 for BBC1, 102 for BBC2, 103 for ITV, and so on, appear to have an order. It is their order of appearance when flicking through with the channel advance button. It could be the historical order in which they started broadcasting, or their transmission frequencies. But these channel numbers could also be seen as relatively arbitrary, and so nominal only. All of these ideas make sense. Many categorical variables in research are not intrinsically either ordinal or nominal. Instead, it depends on how one looks at them.

When a set of categories is clearly intended to be ordinal, such as the order in which competitors finish a sprint race, any ensuing analysis will usually have available the much better and more genuine measures that underlie the order, such as the times taken to finish the race. It is difficult to imagine a situation in which it would be better to analyse using the ordinal data than the evidence used to create that order in the first place. Of course, if there is no good evidence to create the order in the first place (no sprint times, for example) then there is no good order anyway.

Oddly, though, the biggest problem with ordinal values lies in the way they are often treated in research. In psychology especially, but elsewhere as well, questionnaire items are often completed using ordinal scales such as from 'strongly disagree' to 'strongly agree', or similar (and often recorded as 1 for the first category, 2 for the second, and so on). These have been referred to as Likert scales. If something like an attitude or level of agreement cannot be measured with real numbers, and a Likert scale is used, then such a variable is truly categorical ordinal, and must be analysed as such (see Chapter 3). This is not a problem.

However, there is widespread disregard of this consequence, and such ordinal scales are routinely processed as though they were real numbers. They are not real numbers because the intervals along the points in the scale are not equal. We cannot show that the amount of attitude change involved in moving from 'strongly disagree' to 'disagree' is exactly the same as the change involved in moving from 'disagree' to 'unsure' or to 'agree'. In fact, that equivalence seems very unlikely. We cannot do arithmetic with levels of agreement or other variables like this, even if they are expressed numerically. They are not real number measurements, as we have defined these so far.

So when you see research papers and reports using ordinal scales as though they were real numbers (by presenting their mean and standard deviation, for example), then you know that their author does not really know to handle numeric data! Yes, this abuse is still widespread, but that does not make it right. Treat numbers properly and with respect, not sacrificing your honesty for short-term apparent convenience. This does not mean that a Likert scale cannot be used to generate real number **outcomes**, if adapted properly (see examples in this book). Nor does it mean that the simple 'strongly agree' to 'strongly disagree' scale should not be used, as long as it is only used with techniques appropriate for ordinal values (including logistic regression described in Chapter 19).

## Ratio versus interval

One of the aspects of measurement that is routinely proposed by social science researchers is that there are four levels of measurement (Stevens, 1946). These are termed **ratio**, **interval**, ordinal and nominal. We have met three of these already. 'Interval' refers to real number variables with equal intervals on the scale, such as measurements of height in metres, unlike degrees of agreement and disagreement (above).

Ratio variables are also real number variables. Like them, there is a direct analogy between numbers on the scale and properties of the thing being measured. Most significantly, when there are four people in a room this is more than two people in a room; it is exactly twice as many. Twice is a ratio. When someone is 6 standard units tall, they are three times as tall as someone 2 units tall, and so on. Conceptually, a further key point here is that the room can be empty, and the absence of a person will yield a measure of 0 units tall. Some scales can also have negative units, and these work in the mirror-opposite way to the positive units.

Most people encounter such numbering systems in their everyday life, and deal with them perfectly well without knowledge of measurement levels. Many of the figures used in social science are similarly uncomplicated in this regard. These include the number of staff or patients in a hospital, the amount of government funding per pupil at school, or the proportion of Black engineers. All of these figures can be understood without reference to levels of measurement. Of course, many of these measures are more complex and thus error-prone than a simple headcount of people in a room. But it is not their ratio nature that makes them so.

A merely interval measure, on the other hand, is usually described as being like a ratio measure in all respects but without a genuine zero point (such as no people in the room, or no passes in a test). This distinction makes little difference to how we use ratio and interval scales in practice. And there are few non-ratio interval scales in social science. We may add or subtract both ratio and interval numbers (to find that one room is 10 degrees hotter than another, for example), and there are very few analyses that call for only ratio or only interval values.

## Discrete versus continuous

A further distinction, one that again makes little practical difference for social scientists, is between **discrete** and **continuous** measurements (which we have already met under the guise of counts and measures). Many real number measures are continuous in nature. A value for that variable could be anywhere along a continuum. Measurement of height is a reasonably clear example. Within practical extremes, a person's height could be measured as being anywhere on a scale of height.

A discrete variable would be a real number just like a continuous variable, and most of their characteristics are the same. The difference, as the name suggest, is that discrete data can only take on certain values (sometimes, but not always, whole numbers). An example might be the marks awarded to a university student for an assignment. Marks or percentages awarded for student assignments are usually whole numbers such as 72 or 50%.

Like the distinction between ratio and interval variables, discrete or continuous variables will rarely, if ever, make a difference to how you will analyse them using the techniques described in the rest of this book.

## CONCLUSION

This chapter has focused on an important distinction in the use of numbers for research – whether they represent true measurements or not. This distinction creates two types of numbers – categorical variables (nominal and ordinal) and real numbers (counts and measurements). Both are needed, and both are useful. As shown in the remainder of this book, the distinction between them relates to how the different types of numbers are handled in analyses.

## Notes on selected exercises

### Exercise 2.1

a   Types of houses are categorical (presumably nominal, but could be ordinal if arranged in a regular order such as by number of rooms or floors). House prices will be real numbers.
b   Both ethnic group and qualification are categorical. The first is nominal. The second is probably ordinal (as discussed in text).
c   Annual income and test scores both sound like real number variables.

### Exercise 2.2

None will be that easy to measure. But age in years has a clear and valid scale, and most respondents would know their own age. The research problem is whether respondents will tell you. Family income is based on a less clear and valid scale. Does it refer to earned income? Who exactly is a family member? But again the biggest problem, even more than with age, is that people may not want to tell you, and they are less likely to know their family income accurately. Self-esteem is the hardest to measure. It is more theoretical in nature, and is not something that the respondent can tell you directly. It cannot be checked, unlike the other two that could be checked via birth certificates or tax returns, for example.

## Further exercises to try

Find a suitable real-life dataset you are interested in (or use one provided on the website). Try to identify real number and categorical variables. Discuss what makes them so.

Run a frequency analysis for some or all variables. Note the amount of missing data and any oddities or discrepancies. Describe or discuss with others how you would handle each.

## Suggestions for further reading

A chapter on the use and abuse of measurement:
Gorard, S. (2010a) Measuring is more than assigning numbers. In G. Walford,, E. Tucker and M. Viswanathan (eds), *Sage Handbook of Measurement* (pp. 389-408). Thousand Oaks, CA: Sage.

A blog with a longer discussion on the nature of real and categorical data:
https://www.formpl.us/blog/categorical-numerical-data

More on the types of numbers, referred to as 'levels of measurement':
Stevens, S. (1946) On the theory of scales of measurement, *Science*, 103(2684), 677-680.

A short book on how to analyse nominal categorical data:
Reynolds, H. (1977) *Analysis of Nominal Data*. London: Sage.

# Part II

# Basic analyses

# 3
# WORKING WITH ONE VARIABLE

## SUMMARY

The book now turns to the conduct of analyses, using the two kinds of numeric data described so far - categorical (including ordinal) and real number variables. This chapter looks at simple descriptive analyses using only one variable at a time, as opposed to later chapters looking at analysing two or more variables together. Using the methods in this chapter, you could address simple research questions such as:

- How likely are ex-prisoners to reoffend?
- How much do people earn in one region?
- What proportion of the population is in each 'social class'?

The chapter starts by examining the initial preparation of a dataset for analysis, including the question of what to do with so-called 'outliers'. It moves on to creating graphs and summaries for categorical variables, and then to graphs and summaries for real number variables. The chapter concludes with a description of two of the common distributions of numbers (uniform and normal distributions) that you will encounter later in the book, and in your own reading and research.

## DATA PREPARATION AND OUTLIERS

One of the main reasons for conducting a separate variable-by-variable analysis early on in your study is that it helps you get a sense of what your dataset is and what it contains. This is an important stage whether you have collected and coded your own new data, or whether the data has been gathered by someone else (secondary, archive or administrative datasets perhaps, or the exercise datasets available on the book's website, https://study.sagepub.com/gorard, or provided by your tutor).

For example, you could run a simple frequencies report for all variables in the dataset. How to run a frequency analysis is explained in the next section. This would not be intended to be a substantive analysis, as such. Just read through the results. Note where data is missing, or where the cases have peculiar values. Note where categorical variables have an inordinate number of categories, especially categories with very few cases in each.

The actual coding scheme for your data should be set by the decisions you or someone else made when collecting data – in the way the questions were asked and the way these could be answered in a questionnaire, for example. It is generally best to collect any data in the format that you want it to be in for analysis. However, sometimes this is not possible, or you want to revise it, and more often you might be using secondary data that already exists and you had no influence over. See Gorard (2003b) for more on the initial coding and cleaning of primary data.

What to do about the missing data that you spot is covered in detail in Chapter 12. Ideally you will not want any missing data, because it will otherwise both reduce the number of effective cases in your dataset and tend to **bias** your results.

The other main part of cleaning your dataset involves looking for invalid or otherwise inexplicable values. If one of your living respondents is recorded as being 176 years old, then there is a problem. If you have a simple **binary** variable recording 0 if someone is not currently employed, and 1 if they are employed, then a value of –73 for one case is a problem. You can spot such issues by looking at your frequency report for each variable.

Whatever you decide to do on the basis of this must be recorded, and reported appropriately. You might decide, quite properly, to recode all such odd values as missing, and then to treat them as you would any other missing data (Chapter 12). Alternatively, if you can check with the original source of the data or a prior copy of the data, you may be able to find a valid figure that had simply been copied incorrectly, for example. Or, under certain circumstances, you might decide that you are fairly sure what the value should have been. A recorded value of 11 for a binary variable with valid codes of 0 and 1 might just be 1, repeated by mistake when typing (by you or the respondent). But making such assumptions could be dangerous, and if you cannot check this somehow then treating the value as missing might be safer.

More importantly, although you can address all of the individual oddities that your frequency report throws up, it is also worth considering whether any error is a symptom of a wider problem. For example, it may be that an apparent error is a valid entry, but for the next variable in sequence (and one item of data entry has been missed). The response for question 7 may have been coded as the response for question 6, for example. In this very common situation, all entries after the error might also be errors, even if they appear to be valid values (question 8 coded as 7, etc.). Maybe the pages in a questionnaire were stapled in the wrong order, or the person entering the data turned over two pages at once, or skipped a page online. You cannot simply assume that any one problem is isolated. That is why the '11' in the example above might not just be a mistyping of 1.

You may also decide that you want to reduce the number of small categories in one or more variables, now that you know their frequency. This can be done on a theoretical basis, such as dividing an ordinal variable into categories above or below a certain value (combining higher and lower managerial occupations in the standard UK economic classification scale, for example). Or it might be done on a numeric basis, by combining all small groups except one.

If there are empty categories in your coding system, then you can just delete these categories and so simplify your future analyses and presentations. If, for example, there are three types of occupation that the individuals in your sample could have had but no one actually has one of these types, then your further analysis will proceed with only two types. You must still report that there were no cases for one category. If there are entire variables with no variation then the variables themselves can be omitted from further analysis. For example, if everyone in your sample owns a car then proceeding with any further analysis of car ownership is a waste of time.

Once all of this is done, there may still be values for some variables that are extreme or odd – like a 103-year-old in work, or a 15-year-old with a university degree – but this is a different issue than these values being clearly invalid. These values are at least possible. Some commentators and resources label such values **outliers** and suggest their removal lest they somehow distort your findings. This may be partly because, as you will discover in the rest of the book, several well-established techniques still depend on squaring values, and this squaring then exacerbates the outlying nature of 'outliers'.

However, it is better to use the potentially valid data that you have, as it is, and including everything. There are very real dangers once analysts start deleting or amending values in their dataset before analysis, because they do not like the look of them. There is a danger of unconscious bias, or of massaging your results towards what you or others expect. You should never adjust the data without good reason and without making your changes explicit in your reporting. Cleaning up data is good research practice. Falsifying data is cheating. The differences can appear very slight.

Instead, use the data you have and, for interest, conduct an analysis with and without any value you are concerned about, to see what difference it really makes (probably less than you feared, as long as you are working with a decent number of cases).

## Exercise 3.1

Imagine that you are looking at an administrative (secondary) dataset provided to you for your research. It has 98,000 cases (all adults resident in one region) and many variables. You run a frequency analysis for all variables. One variable represents the sex of the respondent, and the coding suggests that a value of 1 represents female, and 2 represents male. Some cases have no value, and some have the value 3. What is going on, and what will you do about it? Another variable represents whether the person is living in state care, but it has an odd format. Most cases have no value given, and only a few have the number 1 recorded. What does this mean, and how will you handle it?

## SUMMARISING CATEGORIES

We turn now to our first simple analysis, assuming that the data has been coded and cleaned as far as possible. Imagine that you are looking at a dataset that includes a variable intended to assess the quality of 170 hospitals. The hospitals have been graded in quality as being one of 'good', 'satisfactory', or 'needs improvement'. This is a categorical ordinal variable, as described in Chapter 2. The order is in the perceived quality of the hospitals. Ignoring for the moment the ever-present and important issues of whether the variable is valid (the conceptual issue of whether we can judge the quality of hospitals in this way), and whether there is any missing data (Chapter 12), how could this single variable be analysed?

You could count up how many hospitals there were in each category, also known as working out the frequency for each category. You could convert these frequencies into percentages to try and make it easier for readers to understand the proportion of hospitals with each grade. Imagine that the figures were as in Table 3.1. The percentages are calculated by taking each of the frequencies, dividing it by the total of 170, and then multiplying by 100. So 113/170 is 0.66470588, which when multiplied by 100 is 66.470588. This has been simplified to approximately 66.5% for the category 'satisfactory'.

**Table 3.1** Frequency of quality grades for 170 hospitals

| Grade | Frequency | Percentage |
| --- | --- | --- |
| Good | 33 | 19.4 |
| Satisfactory | 113 | 66.5 |
| Needs improvement | 24 | 14.1 |
| Total | 170 | 100 |

Around two-thirds (113/170) of all hospitals in the dataset are considered of satisfactory grade. This means that satisfactory is the most frequently occurring or modal category – also known as the **mode**. The mode is a kind of average used with categorical variables. The categories have an obvious hierarchy of desirability (the variable is ordinal categorical), and we can say that the top level occurring in the dataset is 'good' and the worst is 'needs improvement'.

Table 3.1 is a perfectly proper way of presenting the summary results for a categorical variable. It is easy to see what the modal, highest and lowest values are, and we can see the precise frequencies. The table can be created easily using analytical software.

---

In Excel, you would need to list the three categories you want to sum (here 1, 2, 3) in a separate column to the 170 figures. Select three cells in a third column, and in the first of these cells type =Frequency(A1:A170, B1:B3), where A1:A170 is the column of figures, and column B has the numbers 1, 2 and 3. Instead of pressing Enter/Return, hold down the Ctrl and Shift keys and press Enter. The results for each category will appear in the three new cells. Excel will perform most of techniques in Chapters 3-15 (but not Chapters 16-19). Just look them up online. Instructions for Excel are slightly more unwieldy than for SPSS, and are not generally shown in the following chapters.

In SPSS, go to the "Analyze" menu, select the "Descriptive Statistics" sub-menu, select "Frequencies", move the variable Grade to the variables box by selecting Grade and clicking on the right-facing arrow. Click OK. Or you can use the first simple example of SPSS syntax below. Open a syntax box in SPSS, paste these two lines in, and then select Run:

FREQUENCIES VARIABLES=Grade

/ORDER=ANALYSIS.

The resulting output would look like this:

| Statistics | | |
|---|---|---|
| Grade | | |
| N | Valid | 170 |
| | Missing | 0 |

| Value | Frequency | Percent | Valid Percent | Cumulative Percent |
|---|---|---|---|---|
| Good | 33 | 19.4 | 19.4 | 19.4 |
| Average | 113 | 66.5 | 66.5 | 85.9 |
| Needs Improvement | 24 | 14.1 | 14.1 | 100 |
| Total | 170 | 100 | 100 | |

The first part of the output confirms that there are 170 cases in our dataset ($N = 170$), and that all cases have a valid value for the variable Grade. The cases are the hospitals in our sample. None are missing. If there were fewer than 170 valid cases we should try and find out why, and begin the missing data process (Chapter 12). The second part of the output is confusing in having so many columns. Most analytical software has this problem (see Chapter 20 on presenting findings simply). It is better to edit it in Word or similar, to create something like Table 3.1, which contains all of the same relevant information but is easier for the reader to understand.

It is generally useful to draw a graph of any dataset early on, to see what it looks like. The best graph for these results would be a simple **bar chart**, as in Figure 3.1. This graph shows the modal category very clearly, and illustrates the count for each category (although the precise count is usually easier to see in a table).

**Simple Bar Count of Hospital grade**

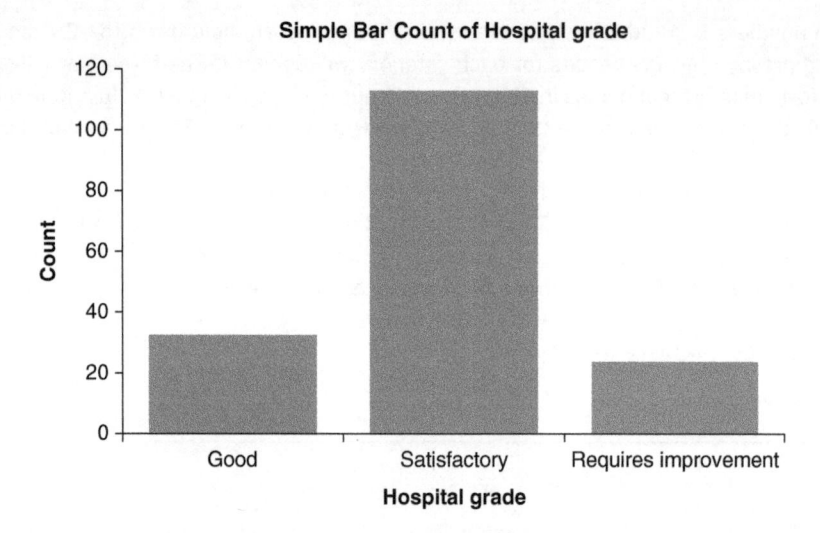

**Figure 3.1** Bar chart of quality grades for 170 hospitals

---

To draw such a graph in SPSS, select the "Graphs" menu, select "Chart Builder", select Bar, drag the first picture in the Gallery to the graph area (top right of the screen), drag Grade to the **x-axis** in the graph area, and click OK.

---

Using only one categorical variable, this is just about the limit of what we can do in terms of analysis – presenting frequencies, percentages and the mode (as an appropriate average), plus the highest and lowest categories (where the categories have a clear order, as here), and a simple graph.

### ▬▬▬ Exercise 3.2 ▬▬▬

Suppose that you were working with a dataset of 200 cases, representing the 200 members of a political assembly. The members come from three different parties. There are 35 from the National Party, 94 from the Liberal Party and 71 from the Socialist Party. What percentage of assembly members are from the National Party? What is the modal party? Is this variable of political allegiance ordinal or simply categorical?

## SUMMARISING REAL NUMBERS

Imagine that another variable in the same dataset was the running cost of each hospital in one area in the past year, measured in pounds. There are 170 numbers. This is a real number variable, as described in Chapter 2. It is a measurement.

One summary of this measurement would be the total for that variable, calculated by simply adding each of the 170 values together. This would represent the total amount of money spent last year on all hospitals in the area. You could also try to find the most frequently occurring value – the mode. But this is less appropriate as an average than it is for categorical data, at least partly because each value may be unique (where no hospital spent exactly the same as any other). Instead, you could find the middle value, if you sort the values into numerical order. This middle value is known at the **median**.

However, the best and most useful average for real numbers is usually the **mean**. This is the total (as above) divided by the number of values (here 170). If the total of all 170 values were £881.33 million, then the mean or average spend for all hospitals would be £881.33 million/170 or around £5.18 million.

The output from analytical software might look like this:

| Descriptive Statistics | | | | | |
|---|---|---|---|---|---|
|  | N | Minimum | Maximum | Mean | Std. Deviation |
| Cost | 170 | 2835746 | 7778376 | 5184312.51 | 1538360.333 |
| Valid N (listwise) | 170 | | | | |

---

To generate the mean in SPSS, go to the "Analyze" menu, select the "Descriptive Statistics" sub-menu, select "Descriptives", move Cost to the variables box, and click OK. Or use the syntax:

DESCRIPTIVES VARIABLES=Cost

/STATISTICS=MEAN STDDEV MIN MAX.

---

The output shows the number of cases again, the lowest and highest cost per hospital, and the mean (average) cost per hospital. On average, each hospital spent just over £5.18 million. The output also shows a value for the '**standard deviation**', as explained in the next section. A better way of presenting the result would be just to state the mean (5.18 million) and standard deviation (1.54 million).

Another way of looking at this data would be to draw a graph, like a **histogram**. Figure 3.2 has no clear shape. It is an example of what is known as a 'uniform distribution', as described in the final section of this chapter.

---

To draw a graph of these costs in SPSS, select the "Graphs" menu, select "Chart Builder", select Histogram, drag the first picture in the Gallery to the graph area (top right of the screen), drag Cost to the x-axis in the graph area, and click OK.

**Simple Histogram of Hospital cost**

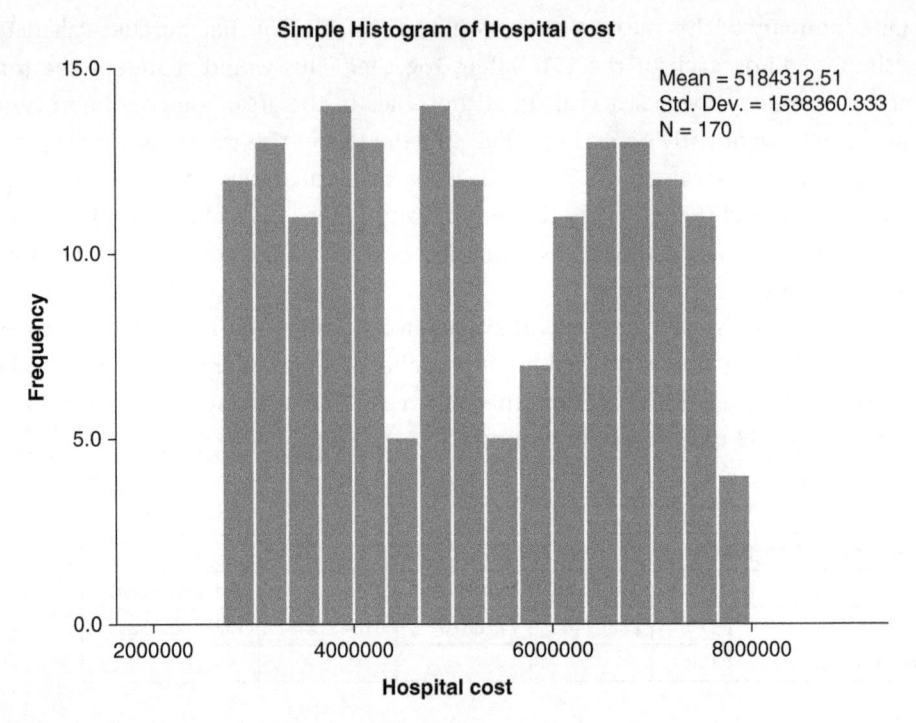

Mean = 5184312.51
Std. Dev. = 1538360.333
N = 170

**Figure 3.2** Histogram of costs for 170 hospitals

## A measure of spread

The second standard bit of simple analysis when working with one real number variable is to assess how spread out the values are around the mean. This is also referred to as the **dispersion** of the scores. Of course we can get a sense of this from a graph showing the distribution of the values (as in Figure 3.2), but we can also summarise this pattern of spread more precisely.

Imagine for this illustration that you have the assignment scores for 10 undergraduate students (second column in Table 3.2). The highest score awarded was 89 and the lowest was 40. The scores add up to 606, and so their mean is 606/10 or 60.6.

**Table 3.2** Assignment scores for 10 students

| Student number | Assignment score | Deviation from mean |
| --- | --- | --- |
| 1 | 47 | −13.6 |
| 2 | 73 | +12.4 |
| 3 | 52 | −8.6 |
| 4 | 51 | −9.6 |
| 5 | 65 | +4.4 |
| 6 | 72 | +11.4 |

| Student number | Assignment score | Deviation from mean |
|---|---|---|
| 7 | 53 | −7.6 |
| 8 | 40 | −20.6 |
| 9 | 89 | +28.4 |
| 10 | 64 | +3.4 |
| Total | 606 | 0 |
| Mean | 60.6 | 0 |

The final column in Table 3.2 shows how far each value is from the overall mean score. In the first such cell the deviation is −13.6 because 47 is 13.6 marks below the mean of 60.6. Around half of the values are above the mean, with positive deviations, and around half are below the mean, with negative deviations. The total of these deviations from the mean score will always be zero by definition. Can you see why?

To create one total figure that summarises these deviations we need to ignore the positive and negative signs in some way. Otherwise the total and mean of all such deviations will always be zero. This is done in Table 3.3 by simply deleting the +/− signs. This is known as using the **absolute value** (or **modulus**) of each number. If we now add these deviations together, their total is 120, and so the mean of all the deviations is 120/10 or 12.

**Table 3.3** Assignment scores for 10 students, with absolute deviations from mean

| Student number | Assignment score | Deviation from mean |
|---|---|---|
| 1 | 47 | 13.6 |
| 2 | 73 | 12.4 |
| 3 | 52 | 8.6 |
| 4 | 51 | 9.6 |
| 5 | 65 | 4.4 |
| 6 | 72 | 11.4 |
| 7 | 53 | 7.6 |
| 8 | 40 | 20.6 |
| 9 | 89 | 28.4 |
| 10 | 64 | 3.4 |
| Total | 606 | 120 |
| Mean | 60.6 | 12 |

So far, we have found the total of all of the values in our variable (assignment scores). We used this to calculate the mean, or the total divided by the number of scores. Using this mean, we then computed how far each individual assignment score is from the mean, ignoring whether they are above or below the mean. We found the total and then the mean of these absolute deviations, yielding a **mean absolute deviation** of 12 for this variable. This shows that on average each individual score was 12 units away from the overall mean score of 60.6.

If the mean absolute deviation is large (perhaps in comparison to the mean), then this shows that the individual scores are very spread out. If the mean absolute deviation is small (in comparison to the mean), then this shows that the individual scores will tend to be close to the mean. If the mean absolute deviation is zero this shows that all of the individual scores are actually the same as the mean (i.e. there is no variation at all in the scores).

The mean absolute deviation is a useful measure of spread in one variable. In real-life research it is said to be preferred to the more complex 'standard deviation' (Barnett and Lewis, 1978; Huber, 1981). It is preferred because it is more efficient in practice, gives each deviation its proportionate place in the result, and is easier for new researchers and others to understand than the standard deviation (explained below). The mean absolute deviation is used routinely in many fields, including astronomy, biology, engineering, information technology, physics, imaging, geography and environmental science (e.g. Amir, 2012; Anand and Narasimha, 2013; Hao et al., 2012).

For a number of reasons that do not concern us here, more than a hundred years ago, it was decided that this mean absolute deviation was too difficult to compute and to represent in algebra (Gorard, 2005). Instead, statisticians suggested what is known as the **standard deviation**. This involves squaring the deviations in Table 3.2, rather than using their absolute values. The squared deviations are shown in Table 3.4. Because squaring involves multiplying each number by itself, we multiply either two positive

Table 3.4 Assignment scores for 10 students, with squared deviations from mean

| Student number | Assignment score | Squared deviation from mean |
| --- | --- | --- |
| 1 | 47 | 184.96 |
| 2 | 73 | 153.76 |
| 3 | 52 | 73.96 |
| 4 | 51 | 92.16 |
| 5 | 65 | 19.36 |
| 6 | 72 | 129.96 |
| 7 | 53 | 57.76 |
| 8 | 40 | 424.36 |
| 9 | 89 | 806.56 |
| 10 | 64 | 11.56 |
| Total | 606 | 1,954.40 |
| Mean | 60.6 | 195.44 |
| Standard deviation | | 13.98 |

numbers giving a positive result, or two negative numbers also giving a positive result (because minus times minus gives plus). The mean of these squared deviations is 195.44. The standard deviation of the 10 assignment scores is defined as the positive square root of the mean of the squared deviations. This is approximately 13.98 (the square root of 195.44) for the figures in Table 3.4.

The standard deviation of 13.98 is slightly larger than the mean absolute deviation of 12. This is because squaring the deviations emphasises the biggest absolute deviations from the mean in either direction, and taking the square root of their sum does not completely overcome this. However, both the mean absolute deviation and the standard deviation are appropriate and largely equivalent methods of summarising how spread out the scores are. And, as shown from Chapter 5 onwards, they are very useful in comparative and investigative analyses. The mean absolute deviation is easier to compute, and has a real-world meaning – it is the mean of the deviations from the overall mean. The standard deviation is harder to compute and to understand, but it is the default version used in most software and other texts. We saw above how the standard deviation is usually generated by analytical software such as SPSS, along with the mean. Use whichever summary of deviations you prefer, or try both!

Even more so than with categorical data, it is worth drawing a graph of real number variables, in order to see how they are patterned, as part of your simple descriptive analysis. In Figure 3.3 the scores are grouped in a histogram showing the frequencies of the assignment scores, in the range 40–50, 50–60, and so on.

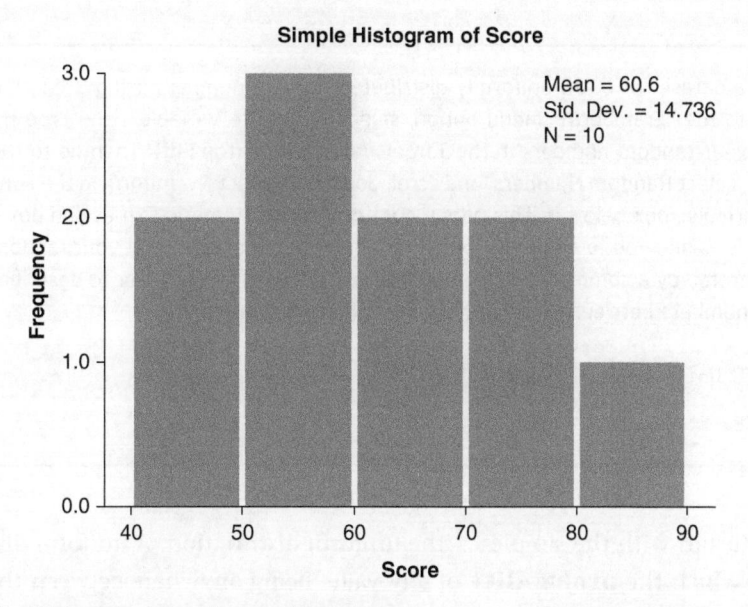

**Figure 3.3** Histogram of assignment scores

Go to "Graphs", select "Chart Builder", select Histogram, choose the first type, drag it to the chart area, drag Score from the variables box to chart area, then click OK.

Drawing a graph, and finding the range, distribution, total, mean and a summary of the deviations from the mean, form the basic set of the analyses that can be conducted with one real number variable.

## Exercise 3.3

You collect six responses from colleagues or other students, and one question concerns their age in years. These ages are given as 27, 32, 21, 44, 23 and 21. What is the total of these figures? What is their mean? What is their absolute mean deviation?

## COMMON DISTRIBUTIONS

This final section of this chapter illustrates two common types of distribution of figures, which are referred to later in the book. A **distribution** shows all of the values in a set of figures (real numbers), in terms of how often each value occurs. Of course, each set of figures will probably be unique in practice, but there are some recognisable *types* of distribution. Both examples below are based on datasets of 1,000 **random** numbers.

To create a dataset of 1,000 uniformly distributed random numbers using SPSS, create 1,000 cases, go to the "Transform" menu option, select "Compute Variable", and type the variable name for your random numbers in the Target Variable box (top left). Then go to the Function group box, select Random Numbers, and scroll down and select Rv.Uniform in the Functions and Special Variables box below it. This puts Rv.Uniform in the Numeric Expression box along with (?,?). This is asking you to enter the lowest and highest values you want your random numbers to be, separated by a comma. This is one of the tasks that is much easier to describe in syntax. To create numbers between 0 and 100 use the following syntax:

COMPUTE Uniform=RV.UNIFORM(0,100).

EXECUTE.

We will start with the simplest – the uniform distribution. A uniform distribution is one in which the **probability** of any value being anywhere between the highest and lowest values is equal. So, for a large number of cases the values will tend to be evenly spread, and the distribution is said to be flat. Figure 3.4 has been generated using Excel to illustrate the usefulness of this software for drawing graphs and simple analysis of data.

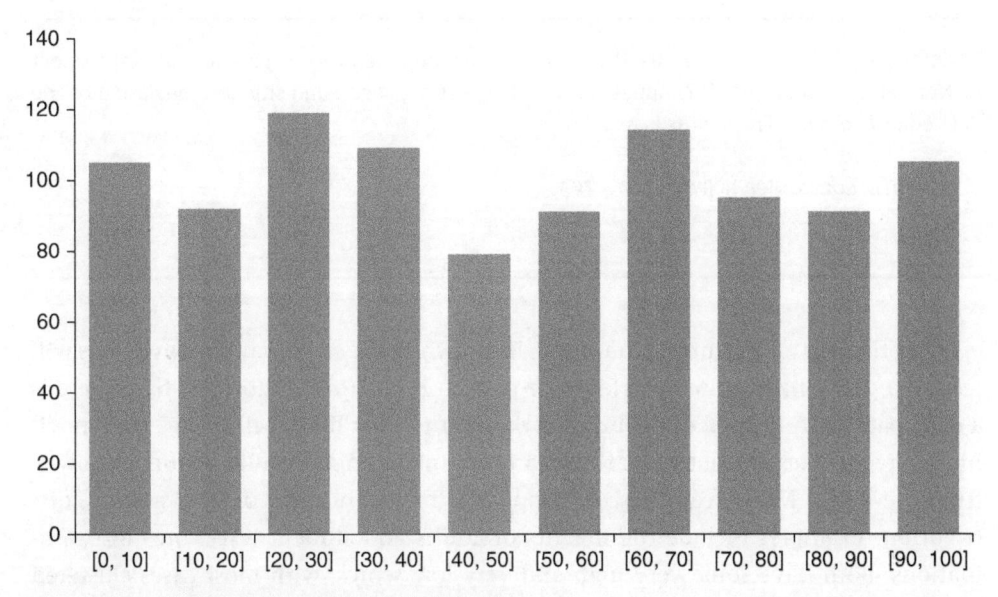

**Figure 3.4** Histogram of uniform distribution

The individual scores have been grouped into bands of scores, here from 0 up to 10, 10 up to 20, and so on. How wide or small these bands are is up to you. We would not want to plot the individual scores because each is likely to occur only once, and we want to get a sense of how clustered they are. These scores are spread out but not completely flat or evenly spread. This is because they are uniform but random numbers. If we ran the randomisation process again we would get different results. We might have most scores in the range 40–50, or any other band. Remember, a uniform distribution of numbers has an equal probability for the values in each band, but this does not mean that there will be exactly equal numbers in each band, on each occasion in practice. That is part of what 'random' means (Chapter 11).

To understand this idea better, think of a pack of playing cards. A pack of cards has an equal number of cards in each of the four suits. This means that dealing a number of well-shuffled cards for one hand of 13 cards leads to a uniform distribution – the probability of spades, hearts, diamonds and clubs in that hand is strictly equal. However, in practice, and in the short term, a hand could have any pattern of suits, even including a hand of only one suit. This does not make the distribution of cards not uniform (or fair). This is an important distinction between mathematical theory and the results of just one randomisation, and will be relevant when we come to look at probabilities in more detail, later in the book.

The second common example is a **normal distribution**. Again, this is generated here using 1,000 random numbers. The software was asked for a mean of 100, and a standard deviation of 20, for these random numbers.

---

The steps would be the same as for the uniform distribution (above). But in the final steps select Rv.Normal, and when the (?,?) appears it is asking for the mean and standard deviation of the desired distribution. The syntax is as follows:

```
COMPUTE Normal=RV.NORMAL(100, 20).
EXECUTE.
```

---

Using the resulting figures, a graph can be drawn in the same way as above. This will lead to a graph like Figure 3.5. Here, the pattern is far from uniform or flat. There is a clear pattern. A normal distribution is thought to look like a bell curve, with much higher frequencies of values near the mean, and increasingly smaller frequencies away from the mean. Many real-life scores appear to have a normal, or near-normal, distribution. Examples include the heights of adults and student test scores in examinations. Both have some very high and very low scores, with most cases clustered around a central average score. Normal distributions will appear again when we consider effect sizes in Chapter 5.

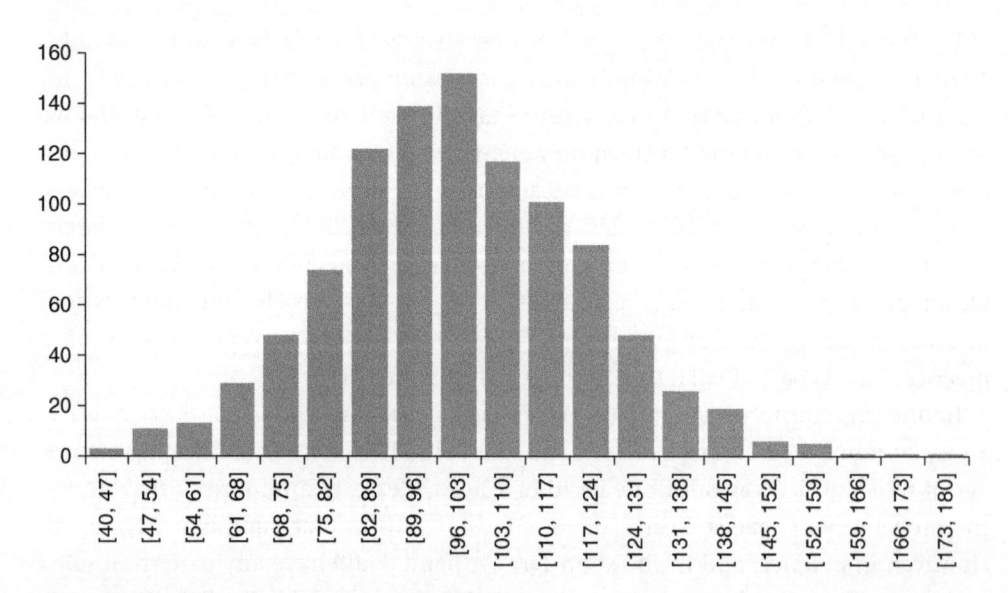

**Figure 3.5** Histogram of normal distribution

The process of creating random number distributions is described here for two reasons. First, the reader can gain confidence in generating random numbers, perhaps for creating random samples (Chapter 10), or allocating cases randomly to groups in an experiment (Chapter 9). Second, the distributions presented here should be considered as near-ideals that you can refer to when examining your own data. It could be important in what follows to know whether your dataset is nearer uniform or normal in shape, or something else. Two further ideal distribution types are shown in Chapter 6.

## CONCLUSION

The univariate analyses described in this chapter are valuable for creating simple findings, and to help you to clean, and make friends with, your data at the outset. The mean, mode and median are all forms of averages, sometimes referred to as measures of **central tendency**. 'Average' is an easier term. The mean absolute deviation and standard deviation are both measures of dispersion around the mean. Another measure of dispersion would be the **inter-quartile range**. The two common distributions, uniform and normal, might be useful to remember, and will be referred to in later chapters. Most of the rest of the book covers more complex analyses, involving several variables at once.

## Notes on selected exercises

### Exercise 3.1

To some extent what the missing values for a variable like sex might mean will depend on the context and the age of the dataset. There is increasing confusion in some social science datasets between what we might term the birth 'sex' characteristics of an individual (sex) and their later self-reported 'gender'. The former is traditionally a binary categorical variable in biology based on reproductive function (Kashimada and Koopman, 2010). The latter refers to traditions and roles associated with each sex, and so perhaps to the power dynamics between men and women (Oakley, 1998). Gender identity refers to how individuals see themselves, in relation to their gender and sex.

The missing values and the value of 3 might be a sign of respondents feeling constricted by the choice they were faced with in a questionnaire, or similar. A small number of people are born with characteristics that are not clearly male or female, and a larger number do not identify as either male or female in self-reports. Of course, it is best to be clear what you are asking for when collecting the data, and the terms 'sex' and 'gender' are used as above in this book. In the example, you could simply create a third category of 'reported as neither male nor female'. You may prefer other terms, or a more complex classification.

Whether someone is living in state care is considered to be an especially sensitive variable, more so than sex and gender, so any work on this needs to take great care that no one is identifiable. You can check with any available sources, but it seems likely that the label '1' represents a case living in care, and so the other cases are not. You can simply create a binary variable – reported to be living in care (1) or not (0) – or in other words replace the blanks with zeros.

*(Continued)*

### Exercise 3.2

This political allegiance is a simple categorical value with no intrinsic order. There are 35 members from the National Party out of 200. This is equivalent to 17.5% (35/200 × 100). The modal or most frequently occurring value is membership of the Liberal Party (94 members).

### Exercise 3.3

The numbers 27, 32, 21, 44, 23 and 21 total 168. Their mean is the total divided by the number of cases (6), so it is 28. The absolute deviations of each value from the mean are: 1, 4, 7, 16, 5 and 7. The total of these is 40, and so the mean absolute deviation is 6.67 (40/6).

---

## Further exercises to try

Find a dataset you are interested in (or use one from the accompanying website). Identify one categorical variable and run a frequency analysis for it. What is the mode? Identify one real number variable and use your chosen software or a calculator to find its mean and standard deviation.

---

## Suggestions for further reading

This text focuses on handling outliers, for those who want to pursue the topic:
Barnett, V. and Lewis, T. (1978) *Outliers in Statistical Data*. Chichester: John Wiley & Sons.

This article discusses graphs and numeric distributions in more detail:
Anscombe, F. (1973) Graphs in statistical analysis, *The American Statistician*, 27(1), 17–21.

This piece is an argument for the use of the mean deviation rather than, or as well as, the standard deviation:
Gorard, S. (2005) Revisiting a 90-year-old debate: The advantages of the mean deviation, *British Journal of Educational Studies*, 53(4), 417–430.

### Help with SPSS

To prepare you if you want to use the SPSS examples, a succinct account of variables, values, file-handling and other housekeeping activities for SPSS is given in:
IBM (2016) *IBM SPSS Statistics 24 Guide*. ftp://public.dhe.ibm.com/software/analytics/spss/documentation/statistics/24.0/en/client/Manuals/IBM_SPSS_Statistics_Brief_Guide.pdf

Again to prepare you if you are interested, this book is perhaps the best single source on how to use techniques in SPSS, and why. It is much more than a manual. Use any edition for any version of SPSS:
Norušis, M. (2011) *IBM SPSS Statistics Guides*. http://www.norusis.com/

Or see this book:
Pallant, J. (2016) *SPSS Survival Manual*. Maidenhead: Open University Press.

Perhaps even more useful for SPSS beginners, look at this YouTube account. It has brief introductory videos explaining how to set up SPSS, how to use the basic interface, entering and editing data, conducting simple analyses such as frequency counts or correlations, and even basic regression modelling. It has the major advantage of not confusing viewers with considerations about significance testing:
https://www.youtube.com/user/patrickkwhite

# 4
# WORKING WITH TABLES OF CATEGORICAL VARIABLES

## SUMMARY

This chapter presents examples of descriptive analyses based on two (or more) categorical variables, using cross-tabulations of frequencies. It includes an extended example from real research showing that such simple techniques can help produce meaningful and useful results. The chapter also looks at a range of approaches to summarising the patterns in tables more succinctly, including the use of odds ratios. The techniques in this chapter could be used to address research questions such as:

- How strong is the relationship between property ownership and political affiliations?
- Do more females attend university than males?
- Are older people more likely than younger people to be working class?

## WORKING WITH TABLES

Imagine that you are interested in two variables of the kinds discussed in Chapter 2, and want to consider the relationship between them. You first have to identify the type of numbers you are dealing with – real measurements, or categories that can only be expressed as frequencies. In the example that follows the two variables to consider are whether someone lives in the north or south of a country, and whether they voted for party A or party B in the last parliamentary election. Your research questions are whether there is a difference in reported voting patterns between north and south, and, if so, in which direction and by how much.

## Exercise 4.1

Using your knowledge from Chapter 2, in the example above what kind of variable (real or categorical) is area of residence and the party voted for?

---

Both variables are of course categorical, and so we can summarise each in the same way as in Chapter 3, in terms of frequencies and percentages.

Imagine you have 100 respondents with responses for both variables. Here the frequencies and percentages will be the same value because there are exactly 100 cases – but you will not always have it that easy! Imagine that 40 cases overall (or 40%) lived in the north, and 60 in the south, and that 50 cases voted for each party. We can begin to summarise our relationship in what is called a cross-tabulation – really just a table (Table 4.1). We can complete the row and column totals from the information we have so far. In total, 50 cases voted for each party, and 60 lived in the south. Therefore 40 lived in the north.

**Table 4.1** Marginal totals for dataset of residence and voting intentions

|                | Voted for party A | Voted for party B | Total |
|----------------|-------------------|-------------------|-------|
| Lives in north |                   |                   | 40    |
| Lives in south |                   |                   | 60    |
| Total          | 50                | 50                | 100   |

We know that 50% of the cases voted for party A, and that 40% of people live in the north. So if there is no link between where people live and how they vote, then 50% of 40% of cases (20) will be in the first empty cell in the table. In this way we can add to each of the four main (empty) cells the numbers we would expect if there were no relationship between area of residence and political support. Under this assumption, each cell should then contain the row total multiplied by the column total, all divided by the overall total (100). These are the **expected values** – expected if there is no relationship, which is what the research is trying to find out. You will hear this expectation of no relationship referred to as the **null hypothesis**.

The full table of expected values would be as in Table 4.2. Note that once one cell is added, the other three empty cells are easy to compute because the cells must add up to the row and column totals. If the first cell is 20 then the one to the right must be 20 as well to add up to 40 in that row. And the one below must be 30 to add to 50 in that column. See? It is that simple. Technically we can say that this table has one **degree of freedom**, because having just one value in one cell means that all other values in all other cells are fixed by that. Only one cell is free to vary before the values of all the others are set.

**Table 4.2** Expected values for dataset of residence and voting intentions

|  | Voted for party A | Voted for party B | Total |
|---|---|---|---|
| Lives in north | [20] | [20] | 40 |
| Lives in south | [30] | [30] | 60 |
| Total | 50 | 50 | 100 |

Finally, we need to tally up how many of our 100 cases there actually *are* in each cell in our dataset, by counting up how many cases reported living in the north and voting for party A, how many reported living in the south and voting for party A, and so on. Imagine the results for our dataset were as in Table 4.3. In our sample, 13 of the people in the north voted for party A and 27 for party B. The total in the north remains at 40 of course, and there are still 50 voting for each party.

**Table 4.3** Observed values for dataset of residence and voting intentions

|  | Voted for party A | Voted for party B | Total |
|---|---|---|---|
| Lives in north | 13 | 27 | 40 |
| Lives in south | 37 | 23 | 60 |
| Total | 50 | 50 | 100 |

### Exercise 4.2

How far apart are the actual values observed (Table 4.3) from those we would expect if there were no pattern (Table 4.2)? How could we summarise how far apart they are?

---

We could generate the figures in Table 4.3 easily via analytical software.

---

Go to the "Analyze" menu, select "Descriptive Statistics", and then "Crosstabs". Drag the Lives variable to the Row(s) box, and the Vote variable to the Column(s) box. Click OK.

```
CROSSTABS
  /TABLES=Lives BY Vote
  /FORMAT=AVALUE TABLES
  /CELLS=COUNT
  /COUNT ROUND CELL.
```

Other than the table below, the output may also include a title and a case processing summary telling you how many valid and missing cases there are in your dataset. Here, there are 100 cases, and none are missing any values. The rest of the output will look like this:

| Lives * Vote Crosstabulation | | | | |
|---|---|---|---|---|
| Count | | | | |
| | | Vote | | |
| | | Party A | Party B | Total |
| Lives | North | 13 | 27 | 40 |
| | South | 37 | 23 | 60 |
| Total | | 50 | 50 | 100 |

This output was simply edited in Word to create Table 4.3. What Table 4.3 shows is that each of the four main cells has a number that is seven different from what we would expect if there were no relationship at all between where someone lived and how they voted (as in Table 4.2). In our sample, the north favours party B and the south favours party A. We could eliminate the difference between Table 4.2 and Table 4.3 by moving 14 cases in total (or 14%). For example, in Table 4.3 we could move seven from party A south to party A north, and seven from party B north to party B south. This figure of 14 out of 100 cases gives an indication of how far away from evenly spread our actual results are. The issue of what this means is taken up in more detail in Chapter 8.

It might be easier, when the numbers involved are larger or more complex (see below), to have the results in percentages rather than just simple frequencies (see Chapter 3). Analytical software will allow you to specify either frequencies or percentages, or both.

---

Conduct the cross-tabulation as above but, before clicking OK, click the Cells button in the dialogue box, and select either or both of Row and Column percentages. In the example, we use row percentages.

    CROSSTABS
    /TABLES=Vote BY Lives
    /FORMAT=AVALUE TABLES
    /CELLS=COUNT ROW
    /COUNT ROUND CELL.

The output looks like this:

| Vote * Lives Crosstabulation | | | | | |
|---|---|---|---|---|---|
| | | | Lives | | |
| | | | North | South | Total |
| Vote | Party A | Count | 13 | 37 | 50 |
| | | % within Vote | 26.0% | 74.0% | 100.0% |
| | Party B | Count | 27 | 23 | 50 |
| | | % within Vote | 54.0% | 46.0% | 100.0% |
| Total | | Count | 40 | 60 | 100 |
| | | % within Vote | 40.0% | 60.0% | 100.0% |

With larger tables this can get quite confusing (having both frequencies and percentages). It is easier to see the patterns in the much simplified Table 4.4. This shows only the percentages in each area voting for each party. These are row percentages, or percentages by row, showing what proportion of people in each area voted for each party. The rows sum to 100%. A column percentage would show what percentage of people voting for each party live in each area (e.g. 33%, or 13/40, of those living in the north voted for party A, and so on). Either way, it is immediately clear that residents of the north strongly favour party B, whereas those in the south slightly favour party A.

Table 4.4 Percentages for dataset of residence and voting intentions

| | Voted for party A | Voted for party B |
|---|---|---|
| Lives in north | 26 | 74 |
| Lives in south | 54 | 46 |

N = 100

## ADDING A THIRD CATEGORY

So far, each of the two variables in our analysis has had only two categories, making the analysis as simple as possible (with four data cells in a two-by-two table). Adding a third category to one variable makes the situation slightly more complex, but the same basic ideas apply. Imagine that 20 of the 100 respondents in our example did not vote or that their vote is unknown. All else remains the same, with 50% of the 80 who did vote supporting each party. The empty table will now be as in Table 4.5.

Table 4.5 Marginal totals for dataset of residence and voting intentions, with missing values

| | Voted for party A | Voted for party B | Did not vote/missing | Total |
|---|---|---|---|---|
| Lives in north | | | | 40 |
| Lives in south | | | | 60 |
| Total | 40 | 40 | 20 | 100 |

Again, we can work out the expected values in each cell, assuming for the present that voting patterns (including not voting) are not related to area of residence. We can multiply the first row total (40) by the first column total (now 40) and divide the result by the overall total (100). This gives the answer 16, and we can also work out that the cell below that must contain 24 in order to total 40 for that column (Table 4.6). But unlike in Table 4.3 we cannot then immediately complete the other four cells by subtracting 16 or 24 from the row totals. This table has two degrees of freedom, meaning that it requires two complete cells in one row before all of the other values fall into place.

**Table 4.6** Incomplete expected values for dataset of residence and voting intentions, with missing values

|                | Voted for party A | Voted for party B | Did not vote/missing | Total |
|----------------|-------------------|-------------------|----------------------|-------|
| Lives in north | [16]              |                   |                      | 40    |
| Lives in south | [24]              |                   |                      | 60    |
| Total          | 40                | 40                | 20                   | 100   |

If we calculate the expected value for the first cell in the party B column, it is again 16. And now the other cells can be filled in by simple subtraction (40 − 16 is 24, 40 − 32 is 8, and 60 − 48 is 12). Table 4.7 is still based on the assumption that the responses are unrelated to area of residence, and it provides a template to compare the actual results with. Of course, we would not really expect our achieved data to match this exactly, but the scale of any difference gives us an indication of how far our sample is from this even spread.

**Table 4.7** Complete expected values for dataset of residence and voting intentions, with missing values

|                | Voted for party A | Voted for party B | Did not vote/missing | Total |
|----------------|-------------------|-------------------|----------------------|-------|
| Lives in north | [16]              | [16]              | [8]                  | 40    |
| Lives in south | [24]              | [24]              | [12]                 | 60    |
| Total          | 40                | 40                | 20                   | 100   |

Imagine that the actual results in our dataset are as in Table 4.8. Is this evidence that there is actually a difference between voting patterns in north and south? Or, put another way, how far are the actual observed results away from the figures in Table 4.7, which are based on assuming that there is no difference in voting patterns? Again, we would have to move 14 cases between cells in Table 4.8 to get the figures in Table 4.7. We could move 6 from party A south to party A north, 4 from party B north to party B south, and 2 from 'did not vote' north to 'did not vote' south. It is still the case that people living in the north seem to favour party B.

**Table 4.8** Observed values for dataset of residence and voting intentions, with missing data

|                | Voted for party A | Voted for party B | Did not vote/missing | Total |
|----------------|-------------------|-------------------|----------------------|-------|
| Lives in north | 10                | 20                | 10                   | 40    |
| Lives in south | 30                | 20                | 10                   | 60    |
| Total          | 40                | 40                | 20                   | 100   |

Performing the cross-tabulation using software is done in exactly the same way as in the simpler two-by-two example above (only the data has changed). Ignoring other output, we get the following result, which can be edited to produce the more helpful Table 4.8. In general, it is a bad idea to present undigested analytical software output in your papers and talks. The job of a table (or graph) is to present the results so that they should make it as easy as possible for readers to see what is meant (Chapter 20).

Lives * Vote2 Crosstabulation
Count

|       |       | Vote2   |         |              |       |
|-------|-------|---------|---------|--------------|-------|
|       |       | Party A | Party B | Did not vote | Total |
| Lives | North | 10      | 20      | 10           | 40    |
|       | South | 30      | 20      | 10           | 60    |
| Total |       | 40      | 40      | 20           | 100   |

Perhaps, if you are a student doing an assignment, you will want to present the full output. In this case you can still describe the results professionally and clearly, but add the full output as an appendix to your main report.

### Exercise 4.3

Wouldn't it be easier in any example like this just to ignore the cases with missing values? The table would be smaller (one degree of freedom again), and easier to explain. If we do not know the party allegiance of some cases, surely they become irrelevant to our analysis?

## COMPARING MORE THAN TWO CATEGORICAL VARIABLES

Using analytical software, it is easy to run more than one cross-tabulation at the same time. Imagine that in our dataset about voting we also have a variable, 'Sex', that records the self-reported sex of each respondent. We could run a cross-tabulation as above, perhaps specifying 'Vote' for the rows, and both 'Lives' and 'Sex' for the columns.

This would produce two separate tables, each like Table 4.4 or 4.8, showing voting patterns by area of residence and by sex. But what if you want to look at all three variables together?

Here you need to specify the variables separately. For example, we could say we want to compare the variables Vote, Lives and Sex as separate dimensions in a table.

---

In SPSS, for example, this can only be done via syntax:

```
CROSSTABS
/TABLES=Lives BY Vote2 BY Sex
/FORMAT=AVALUE TABLES
/CELLS=COUNT
/COUNT ROUND CELL.
```

---

It is not possible to draw up a three-dimensional table on a two-dimensional page, so the output would look something like this (other than the case processing summary).

| Vote2 * Lives * Sex Crosstabulation ||||||
| --- | --- | --- | --- | --- | --- |
| Count ||||||
|  |  |  | Lives || |
| Sex |  |  | North | South | Total |
| Female | Vote2 | Party A | 4 | 14 | 18 |
|  |  | Party B | 12 | 10 | 22 |
|  |  | Did not vote | 4 | 3 | 7 |
|  | Total |  | 20 | 27 | 47 |
| Male | Vote2 | Party A | 6 | 16 | 22 |
|  |  | Party B | 8 | 10 | 18 |
|  |  | Did not vote | 5 | 6 | 11 |
|  | Total |  | 19 | 32 | 51 |
| Any other | Vote2 | Did not vote | 1 | 1 | 2 |
|  | Total |  | 1 | 1 | 2 |
| Total | Vote2 | Party A | 10 | 30 | 40 |
|  |  | Party B | 20 | 20 | 40 |
|  |  | Did not vote | 10 | 10 | 20 |
|  | Total |  | 40 | 60 | 100 |

This is inherently much harder to understand and explain than a two-way comparison, and so this approach should be used rarely. Once you try four-way comparisons (or more) the tables become impossible to read easily (and so are largely unhelpful to you and your reader). It is easier to express these figures as row percentages, and ignore the row totals (they must always be 100% of course). This is still hard to follow (Table 4.9) but is perhaps the easiest way to portray the results, if the focus is on who voted for which party. Drawing simple lines within the table, as here, can help the reader see key differences between the rows.

Table 4.9 Percentages for dataset of residence and voting intentions, with missing data

|  | Lives in north | Lives in south |
| --- | --- | --- |
| Female, voted for party A | 22.2 | 77.8 |
| Female, voted for party B | 54.5 | 45.5 |
| Female, did not vote | 57.1 | 42.9 |
| Male, voted for party A | 27.3 | 72.7 |
| Male, voted for party B | 44.4 | 55.6 |
| Male, did not vote | 45.5 | 54.5 |
| Any other, did not vote | 50.0 | 50.0 |

N = 100

There are only two respondents who reported their sex as other than male or female, and coincidentally neither voted (all of these numbers are actually random). Otherwise, it is clear that both male and female residents were more likely vote for party A if living in the south, whereas it was female voters in the north who were disproportionately voting for party B (54.5%) compared to males (44.4%). However, the north–south difference is more important and more substantial than the male–female one. Any of these observations might be worth further investigation (if this were a real and larger dataset).

## HIDDEN DISADVANTAGE: A REAL EXAMPLE

The next example is based on a real but simple study that involved comparing several categorical variables (Gorard, 2012). In England, children and young people living in families considered to be in poverty, usually on state benefits, are eligible for a free school meal (**FSM**). FSM eligibility data is collected by schools several times a year, and reported to the national Department for Education (DfE). Ignoring the provision of the meal itself, FSM eligibility thus becomes a useful official marker of possible economic and educational disadvantage for pupils at school. It is routinely treated as context for judging both individual and school-level attainment, and as an indicator of school student composition or intake, and has been used as the basis for distributing the additional Pupil Premium funding to schools (Gorard et al., 2019). Knowledge of the quality, reach and limitations of FSM as an indicator is therefore fundamental to accurate decision-making in a number of important areas of policy.

My analysis was based on the Pupil Level Annual Schools Census 2007. It looked at the relationship between different indicators of pupil background and attainment for over 6 million pupils, to help decide how useful FSMs are as an indicator of disadvantage in relation to suggested alternatives. It also looked at how to handle the crucial question of missing data, and described more fully than previously the national picture of who is eligible for FSMs.

Focusing on the categorical variables of FSM eligibility, student mobility (recent arrivals at school or not), living in state care, and some measures of attainment at age 16, reveals some important patterns. The variable for FSM eligibility actually had three values – not eligible (the majority), eligible and not known (around 4% of students in state-funded schools). Using these three categories, it is understandable that the poorer FSM-eligible pupils are more likely to be transient/mobile (perhaps they are Travellers), and more likely to have been living in state care, than the non-eligible pupils (Table 4.10).

**Table 4.10** Percentage of each FSM category with other background characteristics

|  | Not FSM eligible | FSM eligible | Missing FSM code |
|---|---|---|---|
| Student joined in latest 2 academic years | 1.9 | 3.3 | 13.2 |
| Student joined mid-year | 5.2 | 10.7 | 29.4 |
| Been in care while at this school | 0.7 | 0.8 | 3.2 |

What is perhaps less obvious is that pupils whose FSM-eligibility status is not known, in the final column, are far more likely even than the poorer FSM pupils to have these characteristics. FSM-eligible pupils are 14% (or (0.8 – 0.7)/0.7) more likely to have lived in state care than the non-eligible pupils. But pupils missing an FSM code are a staggering 357% (or (3.2 – 0.7)/0.7) more likely to have lived in care than non-eligible pupils, and 300% more likely than even the apparently disadvantaged FSM-eligible group. Missing data for this variable seems to be a marker of extreme disadvantage. It may be that this group includes refugees and asylum seekers as well as Travellers, all of whom are less likely to have the documentation necessary to verify a claim for FSM.

Turning to categorical student attainment outcomes, a similar but inverse pattern applies (Table 4.11). Around 63% of the majority pupils who are not FSM eligible gained the equivalent of five or more high-grade GCSE qualifications at age 16, compared to only 36% for FSM-eligible pupils. This is a key qualification level for continued study and employment. The majority group of pupils are 75% (or (63 – 36)/36) more likely to reach this key level of qualification than their poorer counterparts. However, the majority are a massive 350% (or (63 – 14)/14) more likely to gain this level than those with missing FSM-eligibility codes. And even FSM-eligible pupils, the supposedly disadvantaged group, are 157% more likely to get five 'good' GCSEs than those missing an FSM value. Similar but more extreme patterns occur when considering those whose qualifications include both English and maths.

Table 4.11 Percentage attaining each qualification threshold by FSM status

|  | Not FSM eligible | FSM eligible | Missing FSM code |
|---|---|---|---|
| 5+ GCSEs or equivalent graded A*-C | 63 | 36 | 14 |
| Level 2 with English and maths | 49 | 21 | 7 |

The pupils with missing data are not only more disadvantaged than the officially disadvantaged FSM-eligible group, but also far less likely to attain any level of qualification. The lack of knowledge about this 4% of pupils then causes major, but until recently unnoticed, problems for schools in England. These missing FSM pupils are clustered in particular areas and schools. And the schools have to teach these children and young people who may face considerable challenges (e.g. they are also more likely to have a statement of special educational needs or disability). As shown by Table 4.11, this also means that the schools' average examination results will be lower. But the schools have not (until now) received the extra Pupil Premium funding for most of these pupils (because the schools cannot prove that they are eligible for it!).

In effect, the DfE has treated these pupils as not eligible, and so as not disadvantaged. This means that schools do not receive the extra funding, and these pupils are not counted as context when examining school performance or when the state inspection service Ofsted calls. The schools with lots of pupils missing this key value are trebly disadvantaged – dealing with more hard-to-teach pupils, not getting the funding that they should to provide the resources to deal with it, and then being criticised unfairly for their lower attainment scores.

This was a short, simple study based largely on categorical variables and cross-tabulation. Good analysis does not have to be complex – in fact it hardly ever is. Partly because the findings were so simple, these results were picked up by the media, covered in national TV news, and discussed in Parliament. This led to appropriate calls for changes in how the DfE handled the missing data (see also Chapter 12). Anyone could have done this analysis – undergraduates, master's students or whatever. The data already existed. There must be countless small but genuinely important analyses like this in every field of social science, just waiting for you to have the idea, or to realise the problem.

### Exercise 4.4

Are there datasets relevant to your area of interest that you could use individually, or perhaps by linking two datasets by social group, area or institution (such as a hospital, school or prison)? Can you think of a simple but useful descriptive project like the one above that you could complete efficiently? Discuss with others. There must be lots of ideas waiting to happen. Think, discuss, and then perhaps leave it alone for a bit. Once you have built this foundation, an idea or opportunity often occurs unexpectedly soon after.

## SUMMARISING PATTERNS IN TABLES

This section is slightly more technical, and newer researchers may want to skip it the first time and move on to Chapter 5 and analyses using real numbers.

As the real-life example above demonstrates, summarising the figures and percentages is often enough analysis to permit their meaning to shine out. However, there are numerous other techniques that have been proposed for handling figures in tables that you may come across in some fields of social science.

### Some of the problems

Some researchers and other commentators covering analyses of categorical data routinely present differences between groups over time or place without regard to the scale of the numbers in which the differences appear.

To see why this can be a problem, imagine a social system with only two social classes, in which 1% of social class A attended higher education (HE) while none of social class B did. In this society, therefore, all HE students come from class A, and to be born into class B means no chance at all of education at university. Now imagine a similar society with the same two classes, in which 50% of social class A attend HE, while 49% of social class B do. Would it be true to say that the fairness in opportunities between the two social classes is the same in both societies? Clearly not. But according to policy-makers, most of the media and many academics, both of these societies are precisely equal in their equity of access to HE by social class. Their logic would be that the difference between the participation rates in the first society is 1% (1 – 0), and in the second society it is also 1% (50 – 49).

One of the consequences of this misunderstanding is that other evidence relevant to the topic may then be misinterpreted. If, for example, a researcher or adviser genuinely believed that the figures showed continued inequality of access to HE by social class then they may conclude, wrongly, that a policy or practice that had actually been helpful was unhelpful or harmful. They may then cancel the policy that had been working. The implications of this kind of error, once understood, are remarkable.

Similar issues have appeared in much of social science, including health science (Everitt and Smith, 1979), urban geography (Lieberson, 1981), occupational gender segregation, and social mobility work (Erikson and Goldthorpe, 1991). In each area, absolute differences expressed in simple percentage point terms appear to give a different substantive result than relative differences that take the scale into account (Marshall et al., 1997).

## A solution

A large number of different approaches have been proposed across different fields of study, to try and reconcile these results. Most of the more sensible approaches have many similarities and will provide the same substantive results (Darroch, 1974). In order to compare some of the most common approaches, imagine a generic two-by-two cross-tabulation (Table 4.12). Each cell is lettered from *a* to *d*, for reference. One way of comparing these categories is to use **odds ratios** – see also Chapter 19, as well as Gilbert (1981) and Goldthorpe et al. (1987). The odds for the columns in Table 4.12 (the odds of being in category A if also in category 1, and of being in category B if also in category 2) are defined as $a/c$ and $b/d$. So the ratio of the two odds, or the odds ratio, is $(a/c)/(b/d)$. The odds ratio for the voting example (Table 4.3) would be just under 0.30 (13/27 divided by 37/23), which shows that the odds of voting of for party A if living in the north are less than one-third of the odds for those living in the south.

**Table 4.12** Standard two-by-two table

|  | Category A | Category B |
|---|---|---|
| Category 1 | a | b |
| Category 2 | c | d |

Odds ratios are composition invariant, meaning that if all of the figures were to double (or whatever) the odds ratio would remain the same. This is because $(2a/2c)/(2b/2d)$ is the same as $(a/c)/(b/d)$. The doubling cancels out. Odds ratios are also invariant to changes in only one row (or column). This is because $(2a/c)/(2b/d)$ is the same as $(a/2c)/(b/2d)$, which is the same as $(a/c)/(b/d)$. Odds ratios therefore solve the problem of scale quite simply. When there is no pattern, then $a/c = b/d$ and the odds ratio is 1. When the odds ratio is greater than 1, category A cases are disproportionately in category 1, and so on. Odds ratios are a kind of standardised 'effect' size for categorical variables that could be used to compare two tables over time or place, even if the absolute numbers are on different scales. Effect sizes are discussed further in Chapter 5.

A related alternative is called the 'cross-product ratio'. The cross-products for Table 4.12 would be $a/d$ and $b/c$, and their ratio would be $(a/d)/(b/c)$. This is, in reality, the same as the odds ratio. It is algebraically equivalent.

Another possibility is termed a 'disparity ratio', or 'risk ratio', widely used in health trial evaluation. This is defined as $(a/(a + c)/(b/(b + d))$. It gives a different numeric answer to the other two previous relative approaches, largely because it is answering a slightly different question. It is a measure of the evenness of one characteristic over a number of categories or subgroups, also known as the 'segregation ratio' (Gorard et al., 2003). The segregation ratio is used to study issues such as the stratification of opportunities by social class or ethnic origin.

## CONCLUSION

This chapter illustrated how to summarise the relationship between two categorical variables. Generally, cross-tabulation of frequencies or percentages, with perhaps a graph, is all that is needed to portray any patterns. Where you want to draw comparisons over different contexts such as time or place it is important to consider also changes in scale. The chapter therefore also outlined some of the common approaches to providing effect sizes for such tables.

There are many other available indices and approaches that do similar jobs to those described here (including rho, Yule's Q, segregation indices, and the matching marginals technique). See Gorard and Taylor (2002) and Massey and Denton (1988) for summaries. Although there have been vigorous debates about the merits of each approach, all sensible proportionate approaches provide the same substantive answers. This was illustrated in this chapter, looking at odds ratios and cross-product ratios for example, and is illustrated again in Chapter 5 looking at Cohen, Glass and Hedges effect sizes. Use the simplest approach, if only to make it easier for others to read your research reports.

## Notes on selected exercises

### Exercise 4.1

Both variables are categorical nominal. Any numbers involved in labelling area of residence and voting record are simply used as names.

### Exercise 4.2

This is discussed in the text. The simplest way of summarising how far apart the observed and expected figures are would be to state how much each figure in one table would have to change in order to equal the equivalent figure in the other table. Find the total for these, and perhaps express it as a percentage of all figures. The bigger the answer, the greater the discrepancy between what was observed and what was expected (if there was no pattern).

### Exercise 4.3

It is seldom correct to ignore missing values. Missing or unknown can be as valid a category as any other response, as in the FSM example in this chapter. Ignoring this can lead to heavily biased results. This is such an important topic that Chapters 12 and 13 are devoted to it.

## Further exercises to try

Find a dataset you are interested in (or use one from the accompanying website). Identify two categorical variables and create a cross-tabulation. Work out what the expected table values would have been if there were no pattern, and then work out how far away your result is from that. It helps if you work in the first instance with variables having only two categories.

## Suggestions for further reading

This is a simple explanation of cross-tabulation and the importance of distinguishing between claims about rows and columns. Ignore the stuff about significance tests:
Timpany, G. (2019) *Know your rows and columns.* https://www.cvent.com/en/blog/events/know-rows-columns-cross-tabulation-analysis

Sage has produced a large number of short key texts on particular research approaches. Some are quite old now but remain valuable succinct introductions to each topic. This one takes analysis of categorical data much further, and some of its themes are returned to in Chapter 19:
Hagenaars, J. (1990) *Categorical Longitudinal Data: Log-linear, Panel, Trend and Cohort Analysis.* London: Sage.

An article on the importance of taking scale into account:
Gorard, S. (1999) Keeping a sense of proportion: The 'politician's error' in analysing school outcomes, *British Journal of Educational Studies*, 47(3), 235-246.

This paper looks at a range of methods for summarising tabular and other data, and some of the relationships between them:
Gorard, S. and Taylor, C. (2002) What is segregation? A comparison of measures in terms of strong and weak compositional invariance, *Sociology*, 36(4), 875-895.

# 5
# EXAMINING DIFFERENCES BETWEEN REAL NUMBERS

## SUMMARY

This chapter describes how to look for differences and patterns when comparing the scores for a real number variable (measurements or counts) between two or more categories created by a categorical variable. It also introduces the expression of these differences as effect sizes. Examples of research questions that could be addressed with the techniques in this chapter include:

- How much more do people earn in the south of the country than in the north?
- Do private school students get higher scores at university than state-funded school students?
- Is the reoffending rate for ex-prisoners different between those with a job and others?

## MEASURING DIFFERENCES BETWEEN CATEGORIES

A very common situation faced by social science researchers involves comparing a categorical variable with a real number variable.

---
**Exercise 5.1**
---
In each of the examples in the summary above, identify which variable is the categorical and which is a real number.

Research questions such as those above might be hard to research for practical reasons, such as defining where the north and south end, or getting permission to track ex-prisoners. But once the data has been gathered, such analyses are really rather easy. The simplest approach is to compare the mean (average) score for the real number variable, for each of the groups defined by the categorical variable. The technique is basically the same as described in Chapter 3 for one real number variable. Select only the cases in your data with one value for the categorical variable (e.g. those in the south). Calculate the mean and its standard (or mean) deviation. Now select only those cases in each other category in turn (e.g. those in the north), and calculate the mean and standard (or mean) deviation for each. Put the summary results in a table. In analytical software, this is even easier because it can all be done in one step.

---

Go to the "Analyze" menu, select "Compare Means", and then "Means" (the first item on the submenu). In the dialogue box move Lives to the Independent List box by selecting the variable and then the lower arrow. Move Income to the Dependent list in the same way (using the higher arrow).

MEANS TABLES=Income BY Lives

/CELLS=MEAN COUNT STDDEV.

---

Imagine we are working with 1,000 cases, and 400 of these live in the north as opposed to the south (represented by the categorical variable Lives). The dataset also contains a measure of their income (real number variable Income). Lives is the so-called **independent variable**. This is the categorical variable for whose groups we will be comparing the means. Income is the so-called **dependent variable**. This is the real number variable providing the means for each group. We want to know how much the average income differs between north and south. The output will look something like this:

Means

Case Processing Summary

|  | Cases |  |  |  |  |  |
|---|---|---|---|---|---|---|
|  | Included |  | Excluded |  | Total |  |
|  | N | Percent | N | Percent | N | Percent |
| Income * Lives | 1000 | 100.0% | 0 | 0.0% | 1000 | 100.0% |

Report

Income

| Lives | Mean | N | Std. Deviation |
|---|---|---|---|
| North | 15776.89 | 400 | 7266.497 |
| South | 20286.76 | 600 | 9565.630 |
| Total | 18482.81 | 1000 | 8991.075 |

We can largely ignore the case processing summary. We have no missing cases, and we know there are 1,000 cases, but this is always worth checking for any problems. There are too many decimal places in the figures in the main table (see Chapter 20). With 1,000 cases and self-reported income we cannot claim that the standard deviation is accurate to three decimal places, or five ten-thousandths of a unit, for example. Dealing with this, and cleaned up, the results could be presented as in Table 5.1.

**Table 5.1** Mean earnings by area of residence

|  | Number of cases | Mean earnings | Standard deviation |
| --- | --- | --- | --- |
| Lives in north | 400 | £15,777 | £7,267 |
| Lives in south | 600 | £20,287 | £9,566 |
| Overall | 1,000 | £18,483 | £8,991 |

*Note:* The figures in this table have been **rounded** to whole numbers. So, 15,776.89 is rounded up to 15,777 because 0.89 is greater than or equal to 0.5. But 8,991.075 has been rounded down to 8,991 because 0.075 is less than 0.5.

The answer to our example research question is that, for an imaginary sample of 1,000 cases, people in the north earn £4,510 less than those in the south, on average (20,287 − 15,777). This is a big difference in comparison to the levels of earnings – a difference of around a quarter of average earnings. It could be worth taking note of, reporting, and trying to find an explanation for (Chapter 8).

As in any analysis, it is also useful to see the data on an appropriate graph early on, to help understand what is going on. Here we want to see not the overall distribution (as in Chapter 3), but to examine any differences between the distributions for incomes in the north and south. We want to overlay the two graphs together and see the differences. The output will look like Figure 5.1.

---

Go to the "Graphs" menu, select "Chart Builder", and drag the second (stacked) histogram icon from the choices at the bottom left to the Chart preview area. Now drag Income to the x-axis. Select the Groups/Point ID tab at the bottom left, and tick "Grouping/stacking variable". A stack box (Stack: set colour) appears at the top right of the Graph preview. Drag Lives to this stack box. Click OK.

---

This graph shows that the income for those in the south (the lighter, generally lower, bars) is more variable, as also summarised by the higher standard deviation for the south (£9,566 in Table 5.1) than for the north (£7,266). Because the incomes in the south are more variable, the frequency for any increment on the x-axis is lower than for the north where the incomes are more bunched up. Both sets of scores are limited at the lowest end by zero (incomes are not negative), but some incomes in the south go much higher on the x-axis (which is linked to the higher average for residents in the south).

## 54 | HOW TO MAKE SENSE OF STATISTICS

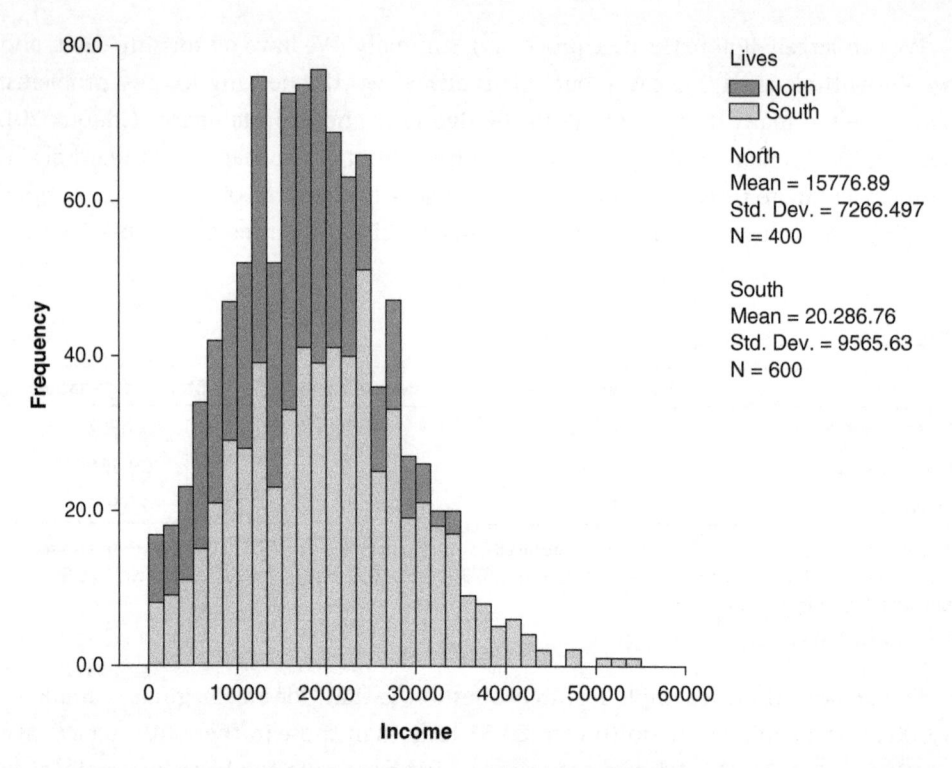

**Figure 5.1** Comparing incomes for residents in north and south

Legend — Lives: North, South

North
Mean = 15776.89
Std. Dev. = 7266.497
N = 400

South
Mean = 20.286.76
Std. Dev. = 9565.63
N = 600

In one way that is the end of the matter. You could leave your analysis here, with the mean difference between the two variables, plus the two means, their standard deviations and a graph.

## STANDARDISING THE DIFFERENCE

However, as with the odds ratios in Chapter 4, we may also want to present the difference in a more standardised summary way, not based on the units of currency of the local income. We could say that the difference between the means of £4,510 is 24.5% of the average of all incomes in the dataset (4,510/18,423). This is a fair summary that takes into account both the scale of the difference and the scale of the figures in which that difference appears. A difference in budgets of £4,510 between two complete countries spending billions is not really a difference at all. The difference between being paid £4,510 and nothing is a colossal difference. Watch out for confusion concerning this in the literature (Gorard, 1999; Gorard et al., 2001). A difference must be considered in relation to the scale of figures that it is a difference in (see Chapter 4).

Our result of a £4,510 difference would mean something very different in an era when the mean earnings were less, and could be trivial in the future if inflation means that the figures for average earnings are much higher. And it does not make any sense to compare the figure 4,510 with a figure from another country based on a completely different currency (although we could convert all figures into a common currency). A general solution to such problems is to create a standard or common version of the difference between the means.

## Proportionate differences

If we do not have the standard deviation or anything like it, then a perfectly acceptable approach is to convert the difference between the means into a proportion between 0 (no difference) and 1 (complete difference because one of the means is zero). This can be done by dividing the difference between the means by their sum. In this example, 4,510 divided by 36,064 (the rounded sum of 15,776.89 and 20,286.76) is approximately 0.125. Or put another way, as we already know, 4,510 is about 25% of the size of overall mean earnings. This method is sometimes used in calculating the 'achievement gap' between different social and ethnic groups of students in education (Gorard, 2000). And it is a fair way of comparing gaps over time and place, because the standard difference is proportionate to the scale (like the currency) which the measures used. If all of the figures in Table 5.1 were 10 times as large, the proportionate difference between north and south would remain the same.

### Exercise 5.2

Try it. Multiply the figures in Table 5.1 by 2, for example. What is the difference between the mean scores divided by their sum now?

### Exercise 5.3

In a sporting match one team scores two points more than the opposing team. So they are the winners. Is the winning margin a comfortable one? Put another way, was this a close match?

A percentage difference or achievement gap as used in the income example is a reasonably acceptable way of presenting a result, and each has been used in real applications such as computing an attainment gap between boys and girls, or between ethnic groups. This approach is particularly useful when we know the scale (how big the numbers are) but not how variable (how spread out) the cases are. However, if we have more information than the plain means of each group, such as the standard deviations or even better the raw scores, then we can present a more careful summary.

We might be able to describe any differences between two sets of numbers as more likely to be scientifically trivial if the numbers themselves show considerable variation over time or place. Or, put the other way around, a difference between two sets of numbers is more convincing if the numbers themselves have a low standard (or mean) deviation. As an imaginary example, consider two sets of 10 numbers:

0, 0, 0, 0, 0, 0, 0, 0, 0, 0

and

2, 2, 2, 2, 2, 2, 2, 2, 2, 2

The sum of the first set is 0 and its mean is also zero. The sum of the second set is 20 and its mean is 2. The standard deviation of both sets is zero, because there is no variability in either set. This makes the difference between the two sets startlingly clear. Every number in the first set (all 0s) is smaller than every number in the second set (all 2s).

Compare this with a different two sets of 10 numbers:

15, 1, 0, –34, 1014, –78, 23, –345, –87, –509

and

15, 1, 0, –34, 1014, –78, 23, –345, –67, –509

The sum of the first set is 0, and its mean is also zero The sum of the second set is 20, and its mean is 2 – both exactly the same as in the first example. The proportionate difference between the means in the income example above would lead us to believe that the difference between the first two sets of numbers is the same size as the difference between the second two sets of numbers. However, everyday common sense tells us that the difference between the first two sets of numbers is quite clear. The same common sense leads us to question the clarity of the difference in the second two sets of numbers. All of the numbers in both of the second sets are the same except for one (–87 or –67). The difference between the two sets (20), appearing in only one number, is much less than the size of some of the numbers in both sets. Formalising this common sense is one of the ways in which an effect size can help.

## Effect sizes

An important way of looking at the scale of any difference therefore lies in relation to the variation in the scores for each group. In the income example above, we have the standard deviation as a summary of the variation in income within each area and the country more generally (Table 5.1). Is the difference between north and south remarkable in comparison to the differences between earnings more generally (within and across north and south)? We have a measure of how spread out the earnings are in north, south and overall – the standard deviations – as well as the means. Using these we can account for both the scale and the variability of our figures when looking at the size of the difference between the scores for each group. We can convert the difference to what is termed an **effect size**. Conversion of the results into a standardised effect size might be done in order to help readers understand the substantive importance of the result, or to allow the result to be synthesised (cautiously) with results from other studies using somewhat different measurements.

Overall, the earnings in our example have a standard deviation (spread) of £8,991. The difference between means (£4,510) is just over half the size of this average spread. In fact the difference divided by the overall standard deviation is 0.502. This measure, of the difference between means divided by their overall standard deviation, is often called an effect size. The result is therefore expressed as a standardised difference, or how many standard deviations apart the two scores are. If the difference between the two means is negative (the second mean is larger than the first) then the effect size would also be negative.

Figure 5.2 shows how an effect size can be envisaged. Two groups of cases are being compared, and the cases in each group are normally distributed with a kind of bell curve (Chapter 3). Each distribution has most cases at or near its mean, and fewer cases as the values get further from that mean. The mean of the left hand distribution is −1, and the mean of the second is +1. Both have a standard deviation of 1. So the effect size of the difference between the two means is 2 (the difference between means divided by the overall standard deviation of 1). If the overall standard deviation is assumed to be 1, this effect size is the same as the difference between means. This value (2) is also the distance between the peaks of the two normal distributions. This is how you can imagine an effect size of this type, as the standardised difference between the peaks of two distributions.

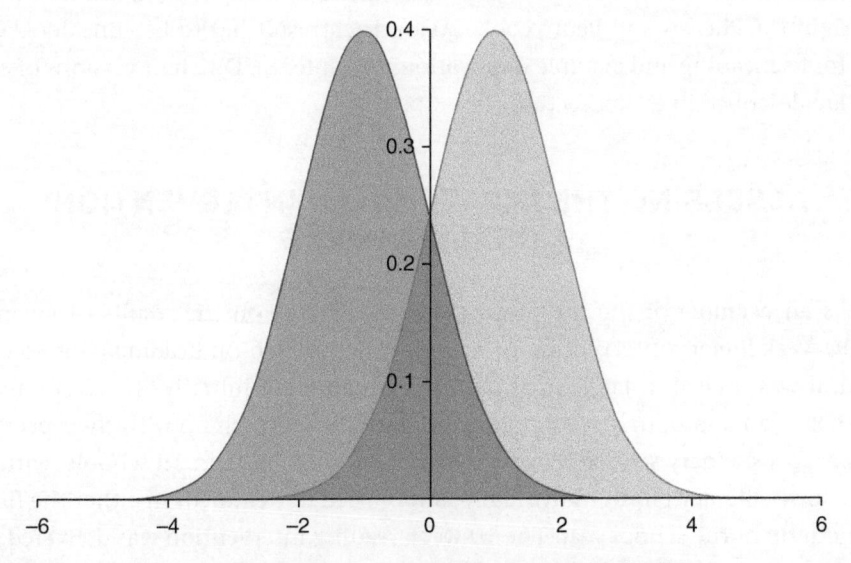

**Figure 5.2** An effect size viewed in terms of two normal distributions

'Effect' size is actually quite a misleading name because it makes it sound as though findings expressed in this way somehow portray a cause and effect situation. In fact, these effect sizes are just standardised differences accounting for both the scale and variability in the numbers involved. It would be better to have another, non-causal, name for it, but the name is now so widely used that this might create confusion.

Because an effect size is a 'standard' score, some commentators have suggested that this makes effect sizes comparable between studies using different measures and approaches. This is not true and can mislead, but there are still scales being published suggesting which effect sizes are worthy of note or represent a substantial difference, and which are not. Perhaps the most often cited scale is for Cohen's $d$ effect size (Cahan and Gamliel, 2011). In this version, an effect size of 0.2 (positive or negative) could be considered small (but just worthy of note), 0.5 would then be a medium effect size, and 0.8 or more a large effect size.

However, Cohen (1977) also pointed out the dangers of using such a scale blindly, or without judgement. These cut-off points are completely arbitrary. Gorard (2006b) describes an example relating to the possible role of aspirin in reducing heart problems, with a tiny fraction as an effect size but which could, if true, prevent deaths for many people for little cost.

We can also look at the stability of a difference over time and place, or of a trend over time. What we might accept as a small difference of no great importance should be considered very differently if it appears annually, for example. A growth in unemployment of 0.1% in a month may be written off as the normal volatility of numbers. A growth of 0.1% for 20 successive months not only adds up to 2% growth, but would be more interesting and worthy of further investigation because it happens so regularly and stably.

Most effect sizes in social science tend to be small, and might seem unimportant at first sight (McPhetres and Pennycook, 2020). Each result has to be considered carefully for its meaning and possible implications (Chapter 8). Different versions of effect sizes are described in a later section.

## ASSESSING THE IMPACT OF AN INTERVENTION: A REAL EXAMPLE

Here is an example of the use of an effect size, based on the results of an intensive 10-week literacy intervention in schools called Switch-on Reading (Gorard et al., 2015). It was trialled in England as part of a government initiative to assist children with below an appropriate level of literacy at age 10 to catch up with their peers, on transfer to secondary school at age 11. Switch-on took place in 19 schools, with 314 Year 7 lower-literacy pupils individually randomised to treatment in either the first or second term of the school year. The 10-week reading intervention was delivered on a one-to-one basis by trained school staff, mostly teaching assistants. The independent evaluation was based on administration of a reading test at the outset and again after one term. Pupil attrition was under 2% (6 out of 314 pupils were missing a test score).

Both randomised groups had very similar scores at the outset (with pre-intervention test means of just over 76), which suggests that the randomisation was effective and so the test of the intervention was fair in that respect (Table 5.2). The groups were comparable, and it is therefore safe to compare the post-intervention scores to discover the possible impact of the intervention on one group after one term (Chapter 9).

Table 5.2  Estimated impact of Switch-on Reading Programme – gain score

| Treatment group | N | Pre-intervention test mean | Post-intervention mean | Standard deviation | 'Effect' size |
|---|---|---|---|---|---|
| Switch-on | 155 | 76.53 | 80.93 | 9.23 | - |
| Control | 153 | 76.14 | 78.73 | 9.29 | - |
| Overall | 308 | 76.33 | 79.84 | 9.26 | +0.24 |

The effect size based on the post-intervention means would be the difference between the post-intervention mean scores of the two groups (80.93 – 78.73, or 2.20), divided by the overall standard deviation (2.20/9.26, or 0.24). It is unlikely that the scores of the missing six pupils (Chapter 12) would have been so divergent between groups that they would have altered the order of magnitude of this effect size (Gorard et al., 2017). The headline finding of this study is therefore that the intervention was effective. The group receiving Switch-on had the same scores as the control at the outset, and did better on the test after the intervention had taken place. Switch-on appeared to improve reading compared to normal teaching. This could be valuable information for schools wanting to help their lower-achieving pupils.

This example show how simple the analysis of an experiment or **randomised control trial** is. You should now be able to conduct the analysis for an experiment, just on the basis of this chapter. As described in Chapter 9, the research design does the difficult work for you. You just need to read off the results and convert them into an effect size.

## ALTERNATIVE VERSIONS OF THE COMMON EFFECT SIZE

As with effect sizes for tables of frequencies (Chapter 4), there have been serious arguments about precisely which version of the common effect size to use. So far in this chapter, and in the Switch-on example, we have looked at the difference between means divided by their overall standard deviations. Cohen suggested using the pooled standard deviations of the two groups. This would involve taking the standard deviation of each group, squaring both, multiplying each by the number of cases in that group, adding the two results, dividing by the total number of cases, and then finding the square root of the result! Sometimes (with tiny samples) $n - 1$ is used rather than $n$, and the total used at the end is $n - 2$.

In the Switch-on example, using post-intervention scores, Cohen's $d$ would be the square root of $(85.19 \times 155 + 86.30 \times 153)/308$. This is the square root of 85.74, or 9.26. This result is exactly the same as the much simpler overall standard deviation that we already used. Cohen's $d$ therefore yields the same effect size of +0.24. It involves all of that pain for so little gain here. Hedges proposed a $g$ effect size that is largely identical to Cohen's (Hedges and Olkin, 1985). The difference involves significance testing, which does not concern us in this book. And anyway this even more complex procedure again leads to the same effect size in practice, unless $n$ is very small.

Glass argued, on the other hand, that we should use only the standard deviation of the control group, who have been unaffected by the intervention in an experimental study (Lenhard and Lenhard, 2016). In the Switch-on example this would be 9.29 instead of 9.26. So Glass's delta effect size would 2.20/9.29, or +0.24 again (to two decimal places). As with so many hotly disputed issues in statistics, the choice makes little or no actual difference to the substantive findings of most research (Gorard et al., 2017). So just use the overall standard deviation. It is not better, but it is far simpler. Or use the mean absolute deviation effect size (Gorard, 2015a). It is even simpler still.

## CONCLUSION

The same approaches as described here can be used for more than two groups. And there are many more different effect sizes that can be used in a variety of contexts or for different kinds of research questions. One used in health for differences between groups is the number needed to treat (NNT). In medical treatment, the NNT is the mean number of people who need to take the treatment for one extra person to benefit (to be cured, report alleviation of symptoms, or whatever), compared to a control group receiving no treatment. There is also a related risk reduction factor in epidemiology. Either or both can be adapted to social science situations.

The $R^2$ value provided by Pearson correlation (Chapter 15) and in multiple linear regression **models** (Chapter 17) is also an effect size estimating how closely related two sets of scores are. Odds ratios used in cross-tabulations (Chapter 4), and in logistic regression models (Chapter 20), are another kind of effect size, appropriate when the numbers involved are frequencies rather than measurements.

Once a standard effect size has been calculated for any study with an intervention, like an experiment or quasi-experiment, it can be compared to the cost, if known, of creating that effect (the number of bangs per buck). We should also try to assess any indirect benefits or dangers. The advantage for those in the treatment group in a study like Switch-on may have deleterious consequences for those not participating in the trial, or there may be value added for all. As stressed throughout this book, computing a numeric answer like an effect size is only the start of the true task of analysis, which is working out what it means. This includes judging whether the effect size is trustworthy.

## Notes on selected exercises

### Exercise 5.1

The categorical variables are the prisoner having a job or not, living in the north or south, and attending a private or state school. The real number scores are the reoffending rate, earnings, and scores at university.

### Exercise 5.2

The results should be exactly the same. Can you see why? This also applies to the percentage difference method, and to the traditional 'effect' size.

### Exercise 5.3

Hopefully, your first thought was to wonder what sport the match was played in. A score of 51 to 49 in basketball denotes a close game. A score of 2 to 0 in football (soccer) represents a reasonably comfortable victory margin.

For example, we could look at the differences in sports scores in terms of the total scores. Thus, a 2-0 football score shows a 100% victory margin, calculated as the difference in scores divided by the sum of the scores, or $(2 - 0)/(2 + 0)$, or 1. The basketball score, on the other hand, could be represented as only a 2% victory margin, or $(51 - 49)/(51 + 49)$, or 0.02. This simple conversion of the raw score difference tells us what we should already know, which is that the football margin of victory is greater. But it does so in a formal and standardised way that can be used to help guard against inadvertent misjudgement in the matter.

## Further exercises to try

Find a dataset you are interested in (or use one from the accompanying website). Identify one categorical variable and one real number. Compute the mean and standard deviation of the real number variable overall, and within each category of the categorical variable. Work out the effect size for the difference in means between the first category and each other category in turn (if there are more than two).

## Suggestions for further reading

It is hard to find good, simple resources on comparing two mean scores, as so many are awash with discussion of significance testing.

*(Continued)*

An article on the importance of using effect sizes and related issues, including why we should not use significance tests for mean differences:

Lipsey, M., Puzio, K., Yun, C., Hebert, M., Steinka-Fry, K., Cole, M., Roberts, M., Anthony, K. and Busick, M. (2012) *Translating the Statistical Representation of the Effects of Education Interventions into More Readily Interpretable Forms*. Washington DC: Institute of Education Sciences.

An article on how to compute effect sizes using the mean (rather than the standard) deviation:

Gorard, S. (2015a) Introducing the mean absolute deviation 'effect' size, *International Journal of Research and Method in Education*, 38(2), 105-114.

Although this new book recommends the use of various effect sizes for presenting summary results for one study, it is important to be aware that effect sizes are not without problems of interpretation - especially when used to try and synthesise the results of many studies together. There is a growing literature on this, of which these papers are two examples:

Morris, P. (2020) Misunderstandings and omissions in textbook accounts of effect sizes, *British Journal of Psychology*, 111(2), 395-410.

Wolf, R., Morrison, J., Inns, A., Slavin, R. and Risman, K. (2020) Average effect sizes in developer-commissioned and independent evaluations, *Journal of Research on Educational Effectiveness*, 13(2), 428-447.

# 6
# SIGNIFICANCE TESTS: HOW TO CONDUCT THEM AND WHAT THEY DO NOT MEAN

### SUMMARY

The book has so far covered everything you will usually need to know when looking at patterns in tables or differences between means. However, if you read social science research involving numbers, and especially if you read training resources for prospective social scientists, you will still come across terms like 'significance test', '*p*-values', '*t*-test', 'ANOVA', and so on. This chapter summarises what such significance tests are, how to conduct them, and what their outputs mean. It also explains why you do not need to use significance tests, why they might harm your findings, and what to do when reading research reports that include such techniques.

This chapter contain fewer exercises, because the techniques in this chapter are not really something you should do, so they are not something you need to practise. Chapter 7 explains in more detail, and more technically, why you should just ignore significance tests in your own work, and when reading the work of others.

### CONDUCTING A SIGNIFICANCE TEST: CHI-SQUARED

Many older resources concerning researching with numbers are based chiefly on the use of significance tests. You will come across these in a variety of formats and using a range of names – including asterisks in table cells (** or *), *p*-values, prob., *t*-tests, chi-squared, ANOVA and *F*-tests. For the writers of such resources, the analysis of numeric data means almost invariably using significance tests. To understand what these significance tests are, before discussing their serious limitations, it is best to work with a few examples.

The first illustration is based on an example from Chapter 4 – the relationship between where people live and how they might vote in an election. Chapter 4 used cross-tabulation, simply tallying how many people lived in the north or south of a country, and voted for party A or party B. In that example there was an obvious pattern. People living in the north voted more for party B and people in the south for party A. Overall, 14% of people would have to change their vote to create balance in the voting patterns between north and south (an odds ratio of 0.30). In a very real sense no more analysis is needed. The next step should be consideration of the meaning of that result (Chapter 8).

However, as an illustration, a basic cross-tabulation can be converted into a test of significance (called chi-squared) using analytical software. Chi-squared is a test of significance used with two (or more) categorical variables, purportedly when trying to decide if they are related. It is based on the difference between the observed and expected values in any table (as shown in Chapter 4).

---

In SPSS, the procedure for chi-square testing is largely the same as for cross-tabulations. Go to the "Analyze" menu, select "Descriptive Statistics", then "Crosstabs". Move the variable Lives (or whatever you are testing) to the Column(s) box, and move the variable Vote (or whatever) to the Row(s) box. This time, before clicking OK, click on the Statistics box on the top right, and then click on the first box, Chi-square. Click OK. Or use the following syntax:

CROSSTABS

/TABLES=Vote BY Lives

/FORMAT=AVALUE TABLES

/STATISTICS=CHISQ

/CELLS=COUNT

/COUNT ROUND CELL.

---

The output will look like this:

| Crosstabs | | | | | | | | |
|---|---|---|---|---|---|---|---|---|
| Case Processing Summary | | | | | | | | |
| | Cases | | | | | | | |
| | Valid | | Missing | | Total | | | |
| | N | Percent | N | Percent | N | Percent | | |
| Vote * Lives | 100 | 100.0% | 0 | 0.0% | 100 | 100.0% | | |

## Vote * Lives Crosstabulation

Count

|      |         | Lives North | Lives South | Total |
|------|---------|-------|-------|-------|
| Vote | Party A | 13    | 37    | 50    |
|      | Party B | 27    | 23    | 50    |
| Total |        | 40    | 60    | 100   |

## Chi-Square Tests

|  | Value | df | Asymptotic Significance (2-sided) | Exact Sig. (2-sided) | Exact Sig. (1-sided) |
|---|---|---|---|---|---|
| Pearson Chi-Square | 8.167[a] | 1 | .004 | | |
| Continuity Correction[b] | 7.042 | 1 | .008 | | |
| Likelihood Ratio | 8.302 | 1 | .004 | | |
| Fisher's Exact Test | | | | .008 | .004 |
| Linear-by-Linear Association | 8.085 | 1 | .004 | | |
| N of Valid Cases | 100 | | | | |

a. 0 cells (0.0%) have expected count less than 5. The minimum expected count is 20.00.
b. Computed only for a 2x2 table

The case processing summary is useful in showing you if there are any missing values, and how many valid cases there are. But it is not part of the test, and will be ignored in this chapter from now on. The next part of the output is simply the cross-tabulation and should be the same as appears in Chapter 4. It shows how many people in each area voted for each party.

The new part is the final table headed 'Chi-Square Tests'. This can be simplified considerably. The footnotes a and b are there to help the reader. They remind us that this output was for a 2 × 2 table (there are two rows and two columns of data in the cross-tabulation because there are two regions and two parties). The notes also provide a summary of the smallest value expected in any cell (calculated by multiplying the marginal totals and dividing by $N$, as described in Chapter 4). The smallest expected cell value in the example is 20, so as the note says there are no cells with expected values less than 5. The row in the table labelled 'Fisher's Exact Test' is only intended for use when the minimum expected cell size is very small, such as less than 5. This does not apply here, and should not be attempted anyway.

The number, $N$, of valid cases is 100, as already shown in the case processing summary and the cross-tabulation. The other rows are slightly different estimates for the 'Value' of chi-squared based on varying assumptions, which similarly do not apply here. Anyway all give similar substantive results. Our focus is only on the first row, reproduced here as Table 6.1.

Table 6.1 Chi-squared test result, voting by area of residence

|  | Value | df | Asymptotic significance (2-sided) |
|---|---|---|---|
| Pearson chi-square | 8.167 | 1 | 0.004 |

The Pearson chi-square value of 8.167 (based on the chi-squared distribution illustrated later in this chapter) is a summary of how much the observed and expected values in the cross-tabulation differ from each other. The label 'df' is short for degrees of freedom, which was discussed in Chapter 4. Here there is one degree of freedom, because once one cell in the 2 × 2 table is filled with a value then the other three cells are defined by the need for each row and column to add up to the row and column totals (treating these as fixed figures). As a rule of thumb, the df is the number of data rows in any table minus 1, multiplied by the number of data columns minus 1.

The idea of df is important because a larger table with more degrees of freedom for values to vary could be expected to have a greater mismatch between observed and expected values just by chance. When assessing the value for the final column in Table 6.1, analytical software takes both the chi-squared (the extent of the mismatch between the observed and expected results) and the df (the amount the cells could vary, by chance) into account.

The final column contains the actual result of the significance test. It is listed as '2-sided' because the probability of a relationship between the values for Lives and Vote was calculated on the basis that people in the north could vote more for party A or more for party B, and the same for people in the south. Two different and opposite relationships are possible. We did not pre-specify that we were looking for a particular difference such as people living in the north preferring party A. That would have been a one-sided prediction.

'Asymptotic' means that the probability given is an estimate.

This estimated probability or $p$-value or significance is 0.004. What does it mean? The value 0.004 is the probability of getting the results in the cross-tabulation (or an even more extreme difference between observed and expected results), if the only reason the observed and expected values differ is by chance, due to the randomisation of the cases involved. Randomisation means that the cases represented in the table were either picked randomly from a larger population (the most likely meaning here), or allocated to their rows (or columns) at random. See Chapter 10 for a more detailed explanation of randomisation.

For this probability of 0.004 to exist there are several other things that must be true – these are called the assumptions of the chi-squared test. First, the true voting patterns of people in the country must be unrelated to where they live (i.e. the same proportion of the *population* must vote for party A in the north as in the south, and so on).

But this is very unlikely to be strictly true in any real country. Second, the cases in the sample must have been selected completely randomly. This means that there can be no missing cases, no cases in the population (such as the homeless) that had no chance of being included, no missing values for cases in the study (e.g. where we know their area of residence but not their vote), and no measurement error or data entry errors. Again, this is very unlikely to be true in any substantial real-life research. Yet it is on these bases that the maths underlying the *p*-value calculation works to generate the likelihood of the results arising solely by chance.

So, a first conclusion is that conducting a significance test, like chi-squared or any other, is hardly ever valid. It is not valid with an opportunity or convenience sample. It is not valid with a strategically selected sample. It cannot be used, and makes no sense, with population figures. It cannot be used with a random sample where there was non-response, replacement, missing data, or errors in data collection. In practice, it is just not valid.

But persevering a little, assuming that we had an ideal random sample, what use is the *p*-value to our research judgement that there is or is not a relationship between areas of residence and voting in our sample? It is not the probability of there being such a relationship (remember, it is based on the assumption that there is no relationship). That would clearly have been a useful result. Instead it is the probability of finding the pattern we found in our sample even if that apparent pattern was a complete fluke. How does that help?

## CONDUCTING A DIFFERENT SIGNIFICANCE TEST: *T*-TEST

Before continuing to examine what a *p*-value is, first consider another example of a significance test – the *t*-test. The *t*-test is used when comparing the differences between the means of two groups (as in the examples in Chapter 5). Using the example of comparing average incomes for those living in the north and the south of the country, the *t*-test is as follows. We want to conduct an Independent-Samples T Test because we are comparing two groups (it is not a One-Sample T Test). It is not a Paired-Sample T Test, which is used when the cases in each group are linked (as, for example, when the scores are for the same cases in each group at different time periods). It is not a One-Way ANOVA or analysis of variance, which comes into play when there are more than two groups.

---

In SPSS go to the "Analyze" menu, select "Compare means" (as in Chapter 10), then pick "Independent-Samples T Test". Move Income (or whatever measure you are assessing for a difference between groups) to the Test Variable(s) box. Move Lives (or whatever) to the Grouping Variable box. This Grouping Variable will appear followed by (?,?), which is requesting the values

*(Continued)*

that define the two groups being compared. Click Define Groups and put 1 for Group 1 and 2 for Group 2, because this is how north and south are coded in the dataset (where 1 is north and 2 is south). Click Continue. Or use the syntax:

```
T-TEST GROUPS=Lives(1 2)
/MISSING=ANALYSIS
/VARIABLES=Income
/CRITERIA=CI(.95).
```

---

The output would be as follows:

### T-Test

### Group Statistics

|  | Lives | N | Mean | Std. Deviation | Std. Error Mean |
|---|---|---|---|---|---|
| Income | North | 40 | 18318.85 | 5974.022 | 944.576 |
|  | South | 60 | 18339.14 | 5338.663 | 689.218 |

### Independent Samples Test

|  |  | Levene's Test for Equality of Variances |  | t-test for Equality of Means |  |  |  |  | 95% Confidence Interval of the Difference |  |
|---|---|---|---|---|---|---|---|---|---|---|
|  |  | F | Sig. | t | df | Sig. (2-tailed) | Mean Difference | Std. Error Difference | Lower | Upper |
| Income | Equal variances assumed | 1.393 | .241 | −.018 | 98 | .986 | −20.283 | 1143.126 | −2288.780 | 2248.214 |
|  | Equal variances not assumed |  |  | −.017 | 77.130 | .986 | −20.283 | 1169.293 | −2348.579 | 2308.014 |

The first part shows the mean and standard deviation of the income variable for each group (north and south). These should be very similar in format to the comparison of means output in Chapter 5. The **standard error** (Std. Error Mean column) is discussed in the next chapter, and need not concern us here. The table shows that the mean income for residents in the south is slightly larger than for the north.

The second table is the output from the *t*-test itself. The idea of a confidence interval is explained in the next chapter, along with the standard error, and these three columns can be ignored here. The **variance** is the square of the standard deviation, or the average of the sum of the squared deviations from the mean. Put another way, the two rows in the second table show the results if the standard deviations in the first table are so similar that they are assumed to be estimates of the same thing, or if that is not true and the variation from the mean differs substantially for each group. Either way, the results here (the significance and mean difference columns) are the same. Ignoring the second row, the simplified output looks like Table 6.2.

**Table 6.2** Result of *t*-test for independent samples – income by area of residence

|  | *t*-value | Degrees of freedom | Significance (2-tailed) |
|---|---|---|---|
| Equal variances of income | −0.018 | 98 | 0.986 |

Degrees of freedom means the same as above and in Chapter 4. With 100 cases, 98 of them would have to be specified before the final 2 had to have a specific value in order to fit with the totals. There are two means. The first is based on 40 cases in the north, and so we could specify 39 cases before the 40th was determined by the overall mean. The second is based on 60 cases in the south, and we could specify 59 values before the 60th must be a specific value in order for the mean to be correct. The degrees of freedom are the sum of 39 and 59, or 98. The *t*-value plays a similar role here to the value of chi-squared. It is based on a theoretical distribution of data to which our data will be compared (see the final section of this chapter).

Again, though, the key figure here is the *p*-value or significance, in the final column. It is said to be **two-tailed** because we did not specify whether we were asking if people in the north earned more (or those in the south did). As with the two-sided chi-squared example, the test is considering both possibilities. It would be a **one-tailed** test if we were concerned with a difference in only one direction.

The probability 0.986 is the likelihood of obtaining the difference in incomes we found (at least 20.283) on the basis that this difference occurred purely by chance. As with the chi-squared test, and all other tests of significance, the assumptions for computing this probability include the complete randomisation of cases, and having no errors in the data. Here the two groups could have been selected independently of each other from the same larger population (two independent samples) or they could represent the entire population of eligible cases, randomised to one or other of two groups (as in an experimental design; see Chapter 8).

## WHAT DOES A *P*-VALUE MEAN?

If we face the likely situation that we have missing cases or data in our dataset, or our cases have not been completely randomised, then it is not possible to compute the probabilities from a significance test (although software will still compute them if we tell it to because the software cannot identify for itself whether a sample is random or not). Instead we could conduct a sensitivity test to see how volatile or robust our results are, which makes no assumptions about randomisation(Chapter 12).

However, under the unrealistic assumptions of complete randomisation of cases, and no errors, the chi-squared test yielded a probability of 0.004, and the *t*-test a probability of 0.986. The first is a rather small probability (near zero chance) while the second represents near certainty (near 1, or 100% chance). The first means that the test result is very unlikely given that it arose solely by chance, and the second means that the result is very likely given that it arose solely by chance. What use are these strange probabilities?

## What a *p*-value does not mean

As illustrated in the two examples above, using a significance test is rarely relevant in real-life research (i.e. its assumptions are rarely, if ever, met), and the peculiar resulting *p*-value is anyway of very limited use to a social science researcher. What we want as analysts is an estimate of the probability that our substantive results arose just by chance. This would allow us to ignore a 'fluke' as a possible explanation for any apparent difference, pattern or trend in our data. We could then move on to considering methodological and substantive/theoretical explanations for our results, as we should do whether we conducted a significance test or not (see Chapter 8).

The problem is that however much we want to know the probability of our results arising by chance, a significance test does *not* provide the probability we want. Put simply, we cannot use the rather useless probability of getting the results we did assuming that they arose by chance as though it were the more desirable and useful probability of our results arising by chance given what the results are. But, among users of significance tests, the confusion between these two very different conditional probabilities is a widespread error.

Think of this analogy to help understand the difference. The likelihood of being a student somewhere in the world if you are reading this book may be quite high (and I hope students do read and enjoy this book!). Put another way, a substantial proportion of the readership of this book (however large or small the readership actually is) could well be students. On the other hand, the likelihood of any specific student somewhere in the world reading this book may be quite small. Put another way, most students in the world will probably never read this, or any, book. This would still be true even if, by the standards of a book on research methods, a relatively large number of people read the book.

So the probability of being a student if reading the book (written more formally as a probability of $S$ given $R$, or $p(S|R)$, may be reasonably large, while the probability of reading the book if you are a student, $p(R|S)$, may be quite small. And we cannot tell from one figure what the other one is. For example, if I heard from the publishers that

62% of people reading the book were thought to be students, this by itself would not tell me what proportion of all students in the world had read the book.

If only 10 students read the book, and no one else, then $p(S|R)$ would be 1, but $p(R|S)$ would be near 0 (10 divided the huge number of students in the world). It should be clear that $p(S|R)$ is different from $p(R|S)$, and one cannot be computed directly from the other. This point is being laboured so that all readers can see that it would be a mistake to assume that because one probability ($p$-value) was small then the other one must be small as well.

Returning to significance tests, it should now also be clear that the probability of obtaining the data in your study given that it arose solely by chance, $p(D|C)$, is not the same as the probability of your data haven arisen solely by chance given what the data is, $p(C|D)$. Therefore if the result of a significance test, or $p(D|C)$, is small (perhaps less than 5%) this does not mean that there is a low probability of the data have arisen by chance. In fact, $p(D|C)$ (the $p$-value) could be near zero, and $p(C|D)$ (what you really want) could still be very large. There is no way of telling one from the other, using only the information provided by significance tests. Significance tests are useless in this respect. In fact, they are worse than useless because they lead to erroneous conclusions (Ioannidis, 2005), vanishing breakthroughs for social science (Matthews, 2002), and harmful real-life consequences for society and individuals. It is safest not to use them, and not to permit yourself to be taken in by them when they are used by others. This issue is picked up again more technically in Chapter 7.

In a sense, all that the complicated figures from a significance test tell us is something we should already know from the scale ($N$) and the effect size of any result. In our voting example, only 33% of those living in the north reported voting for party A, whereas 62% of those living in the south did. This is a substantial difference. We can see this difference without recourse to anything more complicated than the simple comparisons made in Chapter 4. If it genuinely arose by chance then it is an unlikely result. On the other hand, if the income of those in the north and south differs by 20 units, in a figure of around 18,300 (or by around 0.1%), then this difference does not seem very substantial. If this genuinely arose by chance then it is a very likely result. We do not really need to conduct a significance test to understand this trivial fact about either result.

### Exercise 6.1

A colleague of yours has just conducted a significance test with their data (maybe a $t$-test), and reported a $p$-value of 0.01. They claim that this means that their results are unlikely to have been caused by chance alone. In fact the results are so unlikely that your colleague is rejecting their initial assumption that the difference in their data arose only by chance, and is claiming that they have a 'significant' result. Is this the correct interpretation? If not, why not?

## REWRITING RESEARCH RESULTS TO AVOID *P*-VALUE RESULTS

Do not use significance tests in your own writing, unless you have truly randomised cases and want, for some reason, to know $p(D|C)$. Some people might argue that you need to use significance tests in order to get published, because reviewers insist on it. This could happen but it is not a great problem, and it is becoming less of a problem over time. More and more journals across social science do not encourage or even do not allow publication of significance tests.

At time of writing I have over 500 peer-reviewed substantive research articles in journals of many disciplines (education, health, economics, psychology, geography, criminology, sociology, and so on), many of which involve the use of numbers. Every piece I have written has been published, and *none* of them has used significance tests, or anything connected with them. Not using significance tests is no barrier to conducting or publishing research. Just do not use these flawed techniques and your research improves straight away.

What can you do when you are reading someone else's research, and they have used significance tests? In a sense, examples of what to do appear throughout this book. These examples include simply removing references to significance tests and their results from analytical software output or similar, when reporting or examining substantive results. Since the *p*-values and asterisks that you will see provide no reliable guide to the uncertainty in the research results, and may well mislead you and others, it is best to ignore them when reading any research.

This does not mean that the substantive research presenting significance tests, or its methods or conclusions, is necessarily invalid. The research may be invalid, but this cannot be judged through the use or non-use of significance tests alone (Chapter 8). A clear difference, pattern or trend that the research found will still be clear, and a tiny difference will still be tiny. So, when reading, focus more on the design elements (Chapter 9), trustworthiness (Chapter 8), effect sizes (Chapter 5) and sensitivity analyses (Chapter 12). There is more on writing research reports without significance tests in Chapter 21. This whole book is really about how to conduct and read numeric research without using significance tests.

## TWO FURTHER COMMON DISTRIBUTIONS

This chapter ends by illustrating two more common distributions or patterns for data, following the description of the uniform and normal distributions in Chapter 3. The third theoretical distribution to be illustrated is for the chi-squared function, as used in the chi-squared test. The 1,000 random numbers can be created as in both examples in Chapter 3, with minor variations.

# SIGNIFICANCE TESTS: HOW TO CONDUCT THEM AND WHAT THEY DO NOT MEAN | 73

To create 'random' numbers for chi-squared, the random number type is Rv.Chisq, and there is only one ? in the brackets that appear. This ? is replaced by the degrees of freedom, discussed in Chapter 4. Here we use 1 as the degrees of freedom, because that is the same as the example in Chapter 4.

```
COMPUTE Chi=RV.CHISQ(1).
EXECUTE.
```

To create the graph in SPSS, go to "Graphs", then "Chart Builder", select Histogram and drag it to the Chart box, then select the variable (here Chi) and drag to the x-axis, and click OK.

The resulting graph will look like Figure 6.1. This is an exponential decay, with the most commonly occurring values at the lowest end, and rapidly decreasing frequencies as the values increase. You might get a pattern a bit like this if you look at the goals scored by your team in each match over a season in a football league. Many matches would have quite low scores, and a few might have rather higher scores.

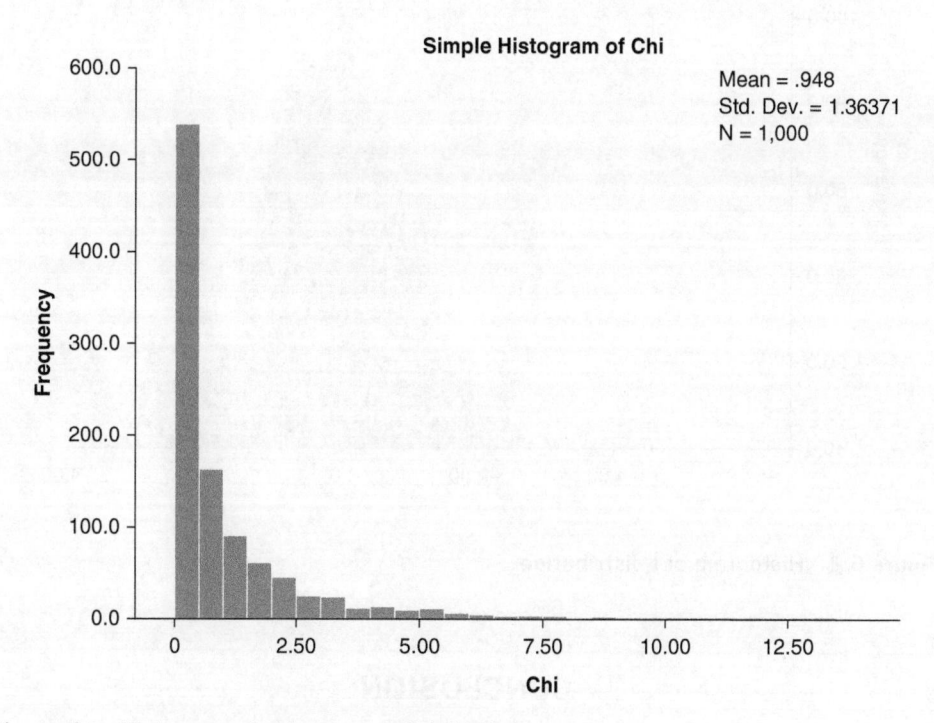

**Figure 6.1** Histogram of chi-squared distribution

The final distribution to be illustrated is the $t$ distribution (or Student's $t$ distribution), as used in the $t$-test.

This can be produced in the same way as for chi-squared (but with the random number function Rv.T in SPSS). Here, the degrees of freedom are selected as 98, because that is the value used in the example above.

```
COMPUTE T=RV.T(98).

EXECUTE.
```

To create the graph in SPSS, go to "Graphs", then "Chart Builder", select Histogram and drag it to the Chart box, then select the variable (here "T"), drag it to the x-axis, and click OK.

A $t$ distribution is something like Figure 6.2. It looks very like a normal distribution. The $t$ distribution tends to be a bit more bunched up, and slightly skewed towards the right, rather than symmetrical, compared to the normal distribution.

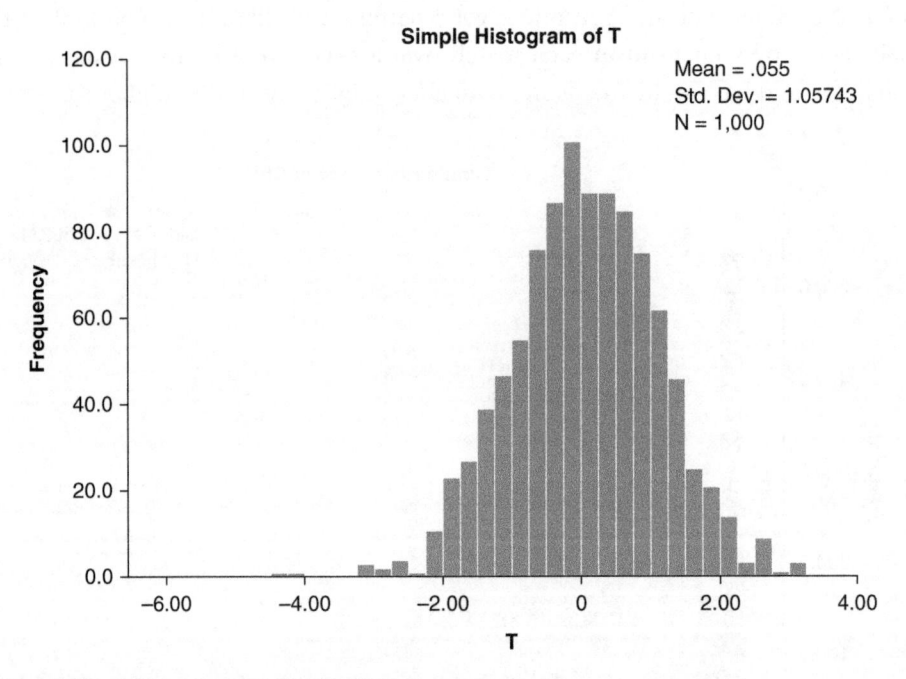

**Figure 6.2** Histogram of $t$ distribution

## CONCLUSION

There are many other tests of significance than the two illustrated here – most notably analysis of variance (ANOVA, or the $F$-test) and its many variants, and tests for ordinal variables and other data formats. If you really want to know more about these then Clegg (1992) describes the steps for some of them, in a simple way, and Siegel (1956) has a clear diagram on the front inside cover showing which of several tests could be appropriate to use in different contexts.

However, all significance tests suffer from the same flaws as described here. They should hardly ever be used, because their assumption of complete randomisation is hardly ever seen in practice. And they provide a probability that is not what most users appear to think it is, which leads to confusion and error. Significance testing also makes research reports using numbers more complicated than they need to be, and so makes some readers either just trust all such research or reject all such research, because they do not understand it.

Significance testing encourages publication bias and the falsification of results (Ioannidis, 2019). The UK Royal Statistical Society and other experts worldwide say that we should stop using the term 'statistically significant', and abandon the idea of null hypothesis significance testing completely. This is to avoid erroneous beliefs and poor decision-making when working with numeric data (Tarran, 2019). Many major journals across health and social sciences have simply banned the publication of significance tests, because authors seem to be unable to heed advice to stop using them (Starbuck, 2016).

Ditch these tests, and your research tends to improve at a stroke, and will permit a much wider readership to begin to judge the trustworthiness of any research that involves numbers. Significance tests make working with numbers complicated, for no gain, and with a very real danger of being misleading. Chapter 7 explains in more detail why this is.

## Notes on selected exercises

### Exercise 6.1

A low *p*-value from any significance test tells you that the outcome you have is unlikely, given that it arose only by chance. This is very different from saying that there is a low probability that your outcome did actually arise by chance. Your colleague is incorrect. They have computed $p(D|C)$. In order to reject the notion that the difference in their data arose only by chance they would need a low probability for $p(C|D)$. But they do not know $p(C|D)$.

Another way of looking at this would be to consider 1,000 different studies, of which we expect a majority not to have a 'significant' difference in their main comparison. Suppose for this illustration that in reality only 4% (40 studies) have a true underlying difference. The researchers in all 1,000 studies conduct a *t*-test. They claim they have found a true underlying difference when their *p*-value is less than 0.05 (5%). So, in the 40 studies with a real difference, 95% (or 38) will be identified accurately, while 2 studies will be missed at the 5% level. In the 960 other studies with no real difference, 912 (95%) will be identified accurately but the other 48 (5% of them) will wrongly appear to be 'significant'. In total, the researchers will identify and report 86 (38+48) significant results, but only 38 out of 86, or 44% of them, will be correctly identified. Using the common 5% significance criterion, in this example

*(Continued)*

most of the studies (56%, which is far more than 5%) will be incorrectly identified as being 'significant'. Significance is terribly misleading.

A key factor routinely ignored in significance testing is the underlying (unconditional) probability of the difference being true in the population being tested. In the example, this base value is 4%. If this figure is changed the percentage of studies identified correctly would also change (Colquhoun, 2014). In reality, we never know this underlying probability – else why would we do the studies? So we cannot tell how inaccurate a significance test is. They are best avoided.

## Suggestions for further reading

A beginner's guide to significance tests, if you really want to pursue this!
Clegg, F. (1992) *Simple Statistics: A Course Book for the Social Sciences*. Cambridge: Cambridge University Press.

How journal editors reject the 'flawed system of null hypothesis testing':
Siegfried, T. (2015). P value ban: Small step for a journal, giant leap for science. *Science News*, 17 March. www.sciencenews.org/blog/context/p-value-ban-small-step-journal-giant-leap-science

More on one of the reasons why journals are banning the use of significance tests:
Colquhoun, D. (2014) An investigation of the false discovery rate and the misinterpretation of p-values, *Royal Society Open Science*, 1, 1-16.

If you have not read this important piece, do so. It begins to explain the 'replication crisis' in social science research:
Ioannidis, J. (2005) Why most published research findings are false, *PLoS Medicine*, 2(8), e124. http://www.ncbi.nlm.nih.gov/pmc/articles/PMC1182327/

# 7

# SIGNIFICANCE TESTS: WHY WE SHOULD NOT REPORT THEM

### SUMMARY

Many texts about analysis with numbers give a lot of space to the process of conducting significance tests. So, for the curious reader, the previous chapter outlined two of these tests, and the ideas and distributions underlying them. This chapter looks in more detail at why we should not use them, and so why significance tests are not presented anywhere else in this book. The chapter is more technical than most others, and can be skipped by those who would prefer to move on to the much more important issue of how to judge the trustworthiness of a research finding (Chapter 8).

### ASSUMPTIONS OF SIGNIFICANCE TESTS, AGAIN

Significance testing, as described in Chapter 6, demands that the data to be analysed comes from a complete random sample. This demand is called a logical premise or an assumption for conducting a significance test. Random sampling and the nature of randomness are covered more fully in Chapter 10. Here it is enough to say that in a true random sample the cases are selected on the basis of chance/luck alone. There can be no structure to the selection of cases, no non-response, no missing cases or data, and no replacement of cases or data (de Vaus, 2002). The mathematical computation underlying a significance test is predicated on this random sampling, as a matter of mathematical necessity, and if the premise is not true then anything that follows from a significance test is automatically wrong or meaningless (Berk and Freedman, 2001; Gorard, 2019a). Without this necessary element of randomisation, significance tests should not be conducted (Glass, 2014). In practice, then, all of the deeper logical problems with significance testing should not matter, because as analysts we so seldom work with completely random samples.

Significance tests, *p*-values, confidence intervals, standard errors and power calculations are all useless in practice for the simple reason that random samples generally do not exist in real research. Yet significance testing still abounds – both in stand-alone form, and even more frequently in the processes of modelling such as regression or factor analysis (Wright, 2003). Analysts are either unaware of, or are ignoring, the mathematical assumptions of what they are doing, and some expert 'statisticians' have been among the worst of these.

Research reports have been published containing significance tests based on birth **cohort studies** and other population data, with non-random samples, or with samples designed to be random but in fact not so because of very poor response rates or high dropout. Almost all such studies then further misuse the sampling theory techniques in the sense that they present the significance test outcomes as though they could handle measurement error. In fact, of course, such tests were only created to address random sampling variation. And if sampling variation is ruled out as an explanation of the difference in scores between two groups, for example, then the analyst still has to consider all of the other possible explanations before deciding that the result is substantively important. Social science could make an overnight improvement simply by deciding to use sampling theory only where it is clearly relevant (in a tiny minority of studies).

## THE INVERSE LOGIC OF TESTS

However, even if you have a true randomised sample, your problems are not over. Significance testing still does not make sense. It does not provide a useful answer for you as an analyst, even when used correctly (Gorard, 2010b).

Imagine that in an ideal mathematical world we have a completely unbiased coin that we can toss so that it lands fairly with the head side up half of the time, and it lands with the tails side up half of the time. We toss the coin once, and it lands head side up. We can say that before tossing the coin the outcome we obtained had a 50% (or half) chance of occurring. If we toss the coin again and get another head we can calculate that there was, at the outset, a 25% chance of two heads in two tosses. This is because each individual head had a 50% chance, so two heads in a row had a 50% times 50% chance. To continue this example, there is 12.5% chance of a third head occurring on the third toss of a perfectly fair coin. This kind of probability calculation is regarded as simple, and is part of standard maths courses for quite young school students. The process is mathematical and the results are necessarily true (because they are tautological in the sense that the probability results are simply another way of defining what we mean by an unbiased coin toss).

# SIGNIFICANCE TESTS: WHY WE SHOULD NOT REPORT THEM

To summarise, if we know that the coin is unbiased, then we can deduce the probabilities of different coin-tossing outcomes with relative ease. This is analogous to what significance tests do. Knowing (or at least assuming) everything relevant about the population in an ideal situation (full randomisation, no measurement error, missing data, and so on), significance tests can compute the probability of the achieved result occurring merely by chance.

Now imagine a very different situation, but one that some people clearly confuse with the first. In real life (not maths) we are provided with a coin and have no way of knowing whether it is biased or not. We toss it once, and get a head. We cannot say what the probability of that occurring was, because we do not know if the coin is biased, by how much, or in which direction. We can toss it three times and get three heads in a row. We still cannot calculate fairly the probability of that event occurring because we do not actually know the probability of heads or tails in even one throw (because we do not know if the coin is biased, or in which direction, or by how much). We cannot use the fact that we got three heads in three throws, by itself, to decide whether the coin is biased, or not. By definition, the outcome itself contains no information about whether the coin is biased, and so cannot tell us anything about it. It is this situation that advocates of significance tests mistakenly imagine that their tests can solve.

**Image 7.1** Heads or tails?

A significance test – a *t*-test, chi-squared test, *F*-test or similar – is a mathematical calculation of the probability of a specific outcome (the observed research data) given that we know that the outcome has been generated randomly, and fairly, from a larger set of already known possible outcomes. Under these conditions, you can work out the probability of obtaining data as large or as different as the data you did actually obtain. This is like the situation of computing the chance of throwing three heads in a row with an unbiased coin. A significance test therefore provides you with a probability (likelihood or *p*-value) of obtaining your specific random outcome (your observed dataset).

Logically, and mathematically, this makes perfect sense. Although your significance test calculation may be a bit harder than with coin tosses, it is not really any harder in principle. It will anyway be performed by a computer for you. But there is little point in doing such a calculation, and generating such a probability, because it is of no use for your research. In real life, you will not know whether your research results will have occurred randomly, and so knowing the probability of the results you would have got if they had been random is not relevant to you. Instead, you will be interested to try and work out, given the results that you obtained, whether they had occurred by chance. This is the same situation as in the second coin example. Getting the answer is just not possible with a significance test, and without much more knowledge about the population.

If you do not know whether a coin is biased then you cannot use the results of any number of coin tosses to decide whether it is biased. Similarly, if you do not know whether your research results have been created by chance then you cannot use your results alone to decide whether they were randomly generated. Yet users of significance tests routinely mistake the first kind of logic for the second. They assume that a low probability of occurrence, if the only factor involved were chance, must mean that there is a low probability that the only factor involved in the results actually is chance. This is a serious and potentially very damaging mistake for researchers to make, and it seems to stem from not thinking carefully enough about what they are doing.

The *p*-value from a significance test means that your result has that probability of occurring by chance alone, assuming that it did arise by chance alone. If we toss an unbiased coin 10 times, the probability of any one sequence of results is 1 in 1,024, or around 0.098% (just under one-tenth of 1%). This could be considered a low probability. But one of the 1,024 possible sequences of results *must* occur. Therefore we cannot use the occurrence of any one sequence to argue that the sequence did not occur randomly. This is true whether the sequence is TTHTHHHTTH (0.098% chance) or HHHHHHHHHH (the same 0.098% chance). To believe otherwise would be a misunderstanding of how probability works.

The problem with significance testing arises because the analysts using them do not actually want to know the probabilities they calculate. They are using one kind of calculation (about the sample, or how likely it is to have three heads, or whatever) incorrectly as though it could answer a completely different kind of question (about the population, or what the three heads tell us about the fairness of the coin). Simple logic shows that this is incorrect (Shadish et al., 2002). A low probability of the difference observed in the sample, given no difference in the population, does not mean that there is a high probability of a difference in the population. The two probabilities in isolation are not calculable from each other.

In a two-sample *t*-test when the analyst is looking at a difference in means between two groups, for example, they ask whether this difference is 'significantly' different from zero (for the population from which the samples were drawn). That is, they want to know, given the two samples they actually achieved, whether both of them yield estimates of the same population mean. A significance test assumes from the outset that what is being tested is true for the population, and so calculates the probability of obtaining a specific finding from the random sample achieved (Siegel, 1956).

If we designate as $H$ the null hypothesis that the two samples are from the same population, and we call the data obtained from the two samples $D$, then what analysts want is the probability of $H$ given $D$. This is usually written as the conditional probability $p(H|D)$. And it is this probability that is implied when rejecting a null hypothesis of no difference between the means. Analysts are saying, in effect, that the difference in the data is large enough (or sustained enough) to reject the idea that both samples come from the same population (i.e. to reject the idea that the observed difference between the two samples is solely due to the vagaries of random sampling).

Unfortunately, a significance test like a *t*-test does *not* calculate this probability, or anything like it. Instead, it calculates $p(D|H)$ – the probability of obtaining the difference between the two samples assuming that they were both drawn from the same population. The conditional probabilities $p(D|H)$ and $p(H|D)$ both contain the same terms but they are very different. To assume that they are somehow the same would be like saying the probability of living in London if one owns a flat is the same as the probability of owning a flat if one is living in London. The two probabilities are not the same. Nor is either directly computable from the other. Knowing that 20% of London residents own a flat, for example, does not tell us, without a lot more information as well, what percentage of flat owners in the world live in London.

The difference between the two conditional probabilities has been stressed here because the point is so important, and the misunderstanding is still relatively widespread.

## SIGNIFICANCE TESTS AND THE POPULATION

The actual computation for a significance test involves no real information about the wider population from which the sample was drawn. This means that the same sample from two very different populations would yield the same *p*-values. A sample mean of 50 from your data would, quite absurdly, produce the same *p*-value if the population mean were really 40, 50, 60, 70, etc. This is because the population value is not known (else there would be no point in conducting the significance test), and the entire calculation is based only on the achieved sample value.

To illustrate the common misunderstanding of this, consider a simplified situation. There is a bag containing 100 well-shuffled balls of identical size, and the balls are known to be of only two colours. A sample of 10 balls is selected at random from the bag. This sample contains 7 red balls and 3 blue balls. The analytical question to be addressed is: how likely is it that this observed difference in the balance of the colours between the two samples is also true of the original 100 balls in each bag? The situation is clearly analogous to many analyses reported in social science research. The bag of balls is the population, from which a sample is selected randomly.

A moment's thought shows that it is not possible to say anything very much about the other 90 balls in the bag. The remaining 90 might all be red or all blue, or any share of red and blue in between. There is no way any probabilistic calculation can be used to work it out. Yet the purpose of a significance test analysis in this situation is to find out, via sampling, something about the balance of colours in the bag. Without knowing what is in the bag there is no way of assessing how improbable it is that the sample has ended up with 7 red balls. Once this impossibility is realised, the pointlessness of significance testing becomes clear.

What a significance test does instead is to make an artificial assumption about what is in the bag. Here the null hypothesis might be that the bag contains 50 balls of each colour at the outset. Knowing this, it becomes relatively easy to calculate the chances of picking 7 reds and 3 blue in a random sample of 10 balls. If this probability is small (traditionally less than 1 in 20, or 0.05) it is customary to claim this as evidence that the bag must have contained an unbalanced set of balls at the outset. This claim is obviously nonsense. The mere assumption of the null hypothesis tells us nothing about what is actually in the bag. For example, imagine that the bag started with 80 red balls and 20 blues. The sample is drawn as above, and contains 7 reds. The significance test approach assumes that there are 50 reds in the bag and calculates a probability of getting 7 in a sample of 10 balls. This probability will clearly be incorrect in reality because the balls are less balanced in fact than the null assumption requires.

Now imagine that the sample is still the same but that the bag had 70 blue balls and only 30 red originally (the reverse situation). The significance test approach again assumes that there are 50 reds in each bag and calculates the same probability of getting 7 red balls and 3 blue. This probability will also be clearly incorrect because the balls are still less balanced than the null assumption requires, but now in the opposite direction. But much more absurdly, this second probability *must* be the same as the first one. Both are calculated in the same way, based on the same assumption. So the significance test would give exactly the same probability of having drawn 7 reds in a random sample of 10 from a bag containing 80% reds as from a bag of 30% reds. This absurdity happens because the test takes no account of the actual proportion of each colour in the population. It cannot, since finding out that balance is supposed to be the purpose of the analysis.

Of course the probability of getting 7 reds from a bag containing 80 reds is different, a priori, from the probability of getting 7 reds from a bag containing 30 reds. But the significance test is conducted post hoc. There is no way of telling what the remaining population is from the achieved sample alone.

Anyone who has spotted this misunderstanding will not use significance testing (Falk and Greenbaum, 1995). No one really wants to know the probabilistic answer the tests actually provide (about the probability of the observed data given the assumption), and the test cannot provide the answer analysts really want (the probability of the assumption being true given the data observed). This conclusion is not new (Harlow et al., 1997). It has been known for a long time, perhaps since their earliest adoption, that significance tests do not work as hoped for, and can be harmful because their results are so widely misinterpreted (Carver, 1978). Significance testing and $p$-values are easily misunderstood, give misleading results about the substantive nature of results, and are best avoided (Lipsey et al., 2012).

## BAYES' THEOREM

**Bayes' theorem** puts the above argument into a more mathematical format. The theorem shows how to convert the conditional probability that significance tests compute, which is $p(D|H)$, into the probability that analysts actually want, which is $p(H|D)$. It is:

$p(H|D) = p(D|H) \times p(H)/p(D)$

This formula requires knowledge of two additional probabilities. These are the unconditional probability of the null hypothesis being true regardless of the data, $p(H)$, and the unconditional probability of getting the data regardless of the hypothesis being true, $p(D)$. So, $p(H)/p(D)$ is the base probability for $p(H|D)$ being true. In the disease example in Chapter 21 the base rate is 1%, and in the notes for Exercise 6.1 (Chapter 6) the base rate used is 4%. In reality, though, we will never know $p(H)$, because if we knew the unconditional probability of the hypothesis being true already we would not be trying to test whether this hypothesis were true or not.

The formula clearly shows that the result of a significance test would need to be multiplied by $p(H)/p(D)$ to create the conditional probability of a hypothesis being true. Therefore, if $p(D|H)$ is small it does not follow that $p(H|D)$ is small. If $p(H)/p(D)$ is very large then $p(H|D)$ is much bigger than $p(D|H)$. If $p(H)/p(D)$ is very small then $p(H|D)$ is much smaller than $p(D|H)$. But a significance test only provides the $p$-value or $p(D|H)$. It does not provide either $p(D)$ or $p(H)$, and so it cannot be used to assess the probability $p(H|D)$, of the hypothesis being true.

## POWER CALCULATIONS

Once we have realised the misleading nature of significance tests, used to try and judge the probability of a set of observations occurring by chance, much else is simplified in working with numbers as well. There is a circular calculation you will come across concerning the 'power' of a sample, used to help decide how large a sample needs to be to detect a given effect size, or how small an effect size could be and still be detectable by a sample of a certain size.

A definition of power for an experimental design is as follows: 'Statistical power is the probability that if the population ES [effect size] is equal to delta, a target we specify, our planned experiment will achieve statistical significance at a stated value of alpha' (Cumming, 2013, p. 17). Here delta is the size of the effect we believe we are looking for, and alpha is the significance level (commonly 5%). This definition makes it clear that power calculations are only relevant when a significance test is being conducted to find the 'result'. Therefore, power calculations would only make any sense if an analyst were planning to conduct a significance test. That is why power calculations, like the significance tests on which they rely, are ignored throughout this book.

None of this removes the responsibility of the analyst to have a large enough sample for their purposes, and to take into account the size and variability of what their research is looking for (Chapter 10).

## CONFIDENCE INTERVALS

Some recent writers, having agreed that significance testing is inappropriate and misleading, have suggested using effect sizes with confidence intervals instead, in what has been termed the 'new statistics' (e.g. Cumming, 2013). This new statistics without significance testing is certainly an important step forward for the field. However, it contains a key contradiction. The problems with confidence intervals are largely the same as those for significance tests. The new statistics needs to reject the use of confidence intervals as well. This would be perhaps the biggest step forward to a superstition-free statistics.

### What is a confidence interval?

A confidence interval (CI) around a mean is expressed as the value of the mean plus or minus an amount. The interval would include any scores in that range. If the mean were 50, the amount for the CI might be 10, so the CI range would be values between 40 (50 − 10) and 60 (50 + 10).

CIs are often invalidly portrayed as a range of values within which a specific population value (such as its mean) is likely to be. So, for example, one might imagine that a 95% CI around a sample mean was a minimum and maximum value between which the population mean is 95% likely to be. This is not what a CI is.

A 95% CI around a measurement of a mean, based on a large true random sample, where the measurements themselves are normally distributed, is calculated as the sample mean plus or minus 1.96 times the standard error of the mean (Peers, 1996). The value of 1.96 comes from the fact that 95% of the area under a normal curve lies within 1.96 standard deviations of its mean (and this value would be different for other CIs, such as 90% or 50%). The CI is adjusted for the sampling error of the sampling distribution, estimated as the **standard error** of the sample mean, which is the standard deviation of the sample mean divided by the square root of the number of cases in the sample. The standard deviation itself is the square root of: the sum of the squared deviation of each score from the mean divided by the number of cases in the sample.

The key assumptions underlying CIs are basically the same as for significance tests. The data must be complete, without error, and the cases must be fully randomised. Imagine, in this ideal and unrealistic situation, taking a very large number of random samples of the same size and quality from one set of population figures. Then compute the CI for each of these imaginary samples (I am not sure how in real life). Around 95% of this huge number of CIs would contain the population mean. Your CI for your one-off sample would be just one of these, and so you do not know if it is one of the 95% or the 5%. In real life you do not have this large number of samples. You have only your data, and one example of an estimate of the mean. A specific sample CI cannot show how close a specific sample mean would be to the population mean. Nor does it follow that the CI for a *specific* sample mean must be one of the imaginary 95% of samples that would contain the population mean.

The ensuing difficulties of comprehension lead to the common errors of using CIs as though they could handle non-random cases, and of interpreting a CI as a range of likelihood within which a desired parameter will fall. Neither is true (Watts, 1991). CIs are therefore dangerous (Matthews, 1998). They are not an estimate of how much confidence to have in the result; nor do they offer a likelihood that the true result will lie within that interval. As traditionally used, CIs are really just significance tests in a more complicated guise.

In practice, of course, no real-life datasets are suitable for use with CIs anyway. This is part of the reason why CIs are not advocated in this book. It would obviously be unnecessary to use them in the situation where the population mean is already known. CIs are not required and do not offer any sensible or comprehensible message for anyone working with population data. And they cannot make up for missing data in datasets because an incomplete dataset is not a random sample of course. Nor do they address things like bias or errors in measurement.

## An example of why CIs do not work

Imagine a population of all of the schoolchildren aged 10 in one region. An ideal sample of 81 schoolchildren is selected at random from this known population, and tested for their attainment in maths. Imagine also that all 81 children took part, that the tests were 100% accurate as an assessment of attainment, and that the 81 scores were normally distributed (Chapter 3). Perhaps the mean attainment score of the 81 children was 50 marks, with a standard deviation of 18. The 95% CI for this made-up result would traditionally be computed as being from 50 − 1.96 × (18/9) to 50 + 1.96 × (18/9), or from 46.08 to 53.92. This is and looks like a small interval, and it would give an analyst advocating the use of CIs reasonable confidence that the sample mean of 50 is a robust estimate of the population mean. But should it? What does all of this complicated calculation really tell us about the proximity of the sample mean to the real population mean?

Imagine further that the average score in maths for the population was actually 75 (although our sample mean was 50). This suggests that the mean and CI calculated for the achieved sample, a mean of 50 with a CI from 46.08 to 53.92, are both considerable underestimates. But the analyst would not know this in practice, because they would not know the population mean (else they would not need CIs). They might conclude, wrongly, that 46.08 to 53.92 is a tight range and that 50 is therefore a good estimate for the population mean.

Imagine now that the population mean was really 40, not 75 (with a sample mean of 50 still). What difference does this make to the CI for the one sample? It makes no difference at all of course, because there is no relationship between the calculation of the sample CI and the actual population mean. In this second example, the sample mean is a bit closer to the population mean (10 points or 20% off), although the researcher would not know this. But the CI is calculated in the same way and gives exactly the same answer as in the former example where the sample mean was much further from the population mean (25 points or 50% off). CIs say nothing about the proximity of any one achieved sample mean to the population mean. The achieved sample mean could be close to the population mean or much larger or smaller than it. There is nothing in the measurements from the sample that can tell us what the true situation is. To imagine that they could is to believe in magic, not science.

Given a random sample with complete response, complete measurement accuracy, and of a known size and standard deviation, the CI is about what happens when repeated samples of the same size are drawn. But if so many repeated random samples were really drawn in practice, then the best estimate of the population mean would be the overall mean for the repeated samples (the process would, in effect, simply provide a larger sample and so a better estimate). The use of CIs could not and would not improve this estimate.

Like significance tests, the calculation of CIs makes no reference to the population, and yet CIs are routinely presented as being an assessment of how close a parameter (like a sample mean) is to the unknown population mean. CIs do not work for the same reasons that significance tests do not work. We cannot use a one-sample estimate to decide whether this one sample is a good representation of the population. CIs are just the estimated mean and standard deviation for a sample of a given size, in a different clothing.

## CONCLUSION

Significance tests provide a $p$-value which is the probability of the data in one achieved sample, based on an assumption (or hypothesis) about the unknown data in the whole population. But the results are then routinely treated in other methods resources as though they were the probability of the assumption being true, given the data in the achieved sample, even though these two probabilities are clearly different. The same confusion occurs with the related concepts of confidence intervals and power calculations. The legitimate mathematical constructs of sampling theory based on the true standard error of the population sampling distribution are not much help in real-life research. This is because in real life the relevant values are only known for one sample and not for other imaginary samples or other members of the population, and because so few researchers have complete random samples anyway.

This chapter further demonstrates that significance tests, as outlined in Chapter 6, are rarely applicable in real-life studies, do not make sense even in their own terms, and do not provide the kind of analytical solutions we want as researchers. Their use is at best pointlessly complex, and at worst misleading and dangerous. Having ignored all of these problems for so long, it is no surprise that health and social sciences have experienced decades of 'vanishing breakthroughs' or findings that cannot be repeated or are never useful in practice (Matthews, 1998).

Ditching significance tests and everything associated with them, as this book advocates, makes researching with numbers easier to follow, and our results more secure. Eventually, significance tests should not be taught or used as part of a general module on researching with numbers. Meanwhile they might be taught merely to try and explain what people have been doing in the past, and what this means and does not mean.

Methods experts in medicine, psychology, sociology and education, the American Psychological Association, American Sociological Association, and other bodies agree. And they now advise against the use of significance tests (Fidler et al., 2004; Siegfried, 2015). The *American Journal of Public Health, Epidemiology, Basic and Applied Psychology* and numerous other journals have banned their publication, including most US medical journals (Starbuck, 2016). According to a top-selling author on statistics for undergraduate psychology students, significance tests are 'flawed because the significance of the test tells us nothing about the null hypothesis' (Field, 2013, p. 76). By simply ignoring significance tests your analyses will improve at a stroke, and your work will be more easily understandable to others.

What most people find hard about statistics is this least important and most insecure part. Drop any thought of using significance tests either when conducting or reading research.

## Suggestions for further reading

An argument that most introductory statistics courses appear difficult to students because the course material does not make sense, logically or mathematically:

Watts, D. (1991) Why is introductory statistics difficult to learn? *The American Statistician*, 45(4), 290-291.

A couple of well-known examples of the arguments since at least the 1950s that we should not use significance tests:

Carver, R. (1978) The case against statistical significance testing, *Harvard Educational Review*, 48, 378-399.

Harlow, L., Mulaik, S. and Steiger, J. (1997) *What If There Were No Significance Tests?* Mahwah, NJ: Lawrence Erlbaum.

A demonstration with data that confidence intervals are not measures of uncertainty, but more like indicators of the stability of findings that can be portrayed more easily and comprehensibly in other ways:

Gorard, S. (2019b) Do we really need confidence intervals in the new statistics? *International Journal of Social Research Methodology*, 22(3), 281-291.

For those interested in Bayesian approaches, this book includes some simple explanations of a Bayesian approach, as well as being a generic introduction to statistics. However, Field now says not only that significance tests are flawed and pointless (see above), but also that using and promoting significance tests in his books was a 'statistical faux pas' (www.methodspace.com/profiles/blogs/top-5-statistical-fax-pas):

Field, A. (2016) *An Adventure in Statistics*. London: Sage.

# Part III
# Advanced issues for analysis

# 8
# THE ROLE OF JUDGEMENT IN ANALYSIS

## SUMMARY

The next part of this book is about the validity of research, and the many factors that need to be taken into account when judging the validity and meaning of an analysis. Other chapters in this book describe the first basic steps in running a univariate, bivariate or multivariate analysis. These technical steps in analyses lead to summaries, tables and graphs. But in themselves such artefacts do not definitively answer any research questions or represent any robust conclusions.

Whatever kind of data you are working with, and whatever techniques you use to compute the summary results, the true nature of any analysis lies in your use of judgement along with the research findings. This requires considerable skill (or craft), and is best approached with great clarity, and with any arguments and assumptions laid out as simply as possible. This chapter provides some guidance on the kind of judgement to be used once the basic descriptive numeric findings are clear. It includes aids on judging how trustworthy (valid) your and other researchers' findings are and, if valid, how those findings might be generalised and interpreted.

## INTRODUCTION

The preceding chapters, and others in the following section of the book, explain and illustrate how to compute a range of numeric statistical summaries, such as cross-tabulations, differences between means, effect sizes and regression models. However, computing these summaries and presenting their figures is only the first step in conducting a full research analysis. They are a way of sifting, depicting and understanding the data – just as when we draw a graph or a table. These steps are not really the analysis.

Whatever the findings are, the next steps are to decide how good these results are, what they mean, how important they are, and what might be done about them. The real analysis is the process of judging whether the graph shows a pattern, and what that pattern portrays. Or whether the difference between two or more mean scores is substantial and robust enough to be worthy of further consideration, and how that difference might be explained.

The true skill of analysis lies in the ability to make such judgements and portray them scrupulously so that others can follow and hopefully agree with your reasoning. These steps are not technical ones, and they are consequently more rewarding to do. This chapter discusses the role of judgement in our analyses. The first part of the chapter is about judging the trustworthiness of a research result in terms of its methods. The second part begins a consideration of what a trustworthy research result might mean. The second part is really only relevant if a finding has been judged reasonably trustworthy. There is little point in worrying about what an invalid result means!

## JUDGING THE TRUSTWORTHINESS OF A RESEARCH FINDING

There are several published protocols for judging the robustness of a research finding. Some are intended for the users of research, like That's a Claim (https://thatsaclaim.org/). Some, like the Weight of Evidence framework from the EPPI-Centre, are more for researchers who want to judge a body of work consisting of many individual studies (Gough, 2007). Some, like the security ratings used by the Education Endowment Foundation (EEF), are intended for use when judging the quality of an individual study based on a randomised control trial (RCT) or similar (https://educationendowmentfoundation.org.uk/help/projects/the-eef-security-rating/). Some, like the Maryland Scientific Methods Scale, are like that of the EEF in being concerned with what works, and like that of the EPPI-Centre in being used to judge systematic reviews of evidence (Madaleno and Waights, n.d.).

The somewhat simpler procedure described in this chapter is used to help judge the quality of individual studies which use any design or address any kind of research question. However, all such procedures have much in common, including their aim of assessing the relative validity of research findings.

The first step in the suggested procedure concerns reporting and comprehension (Chapter 21). Whatever approach is used to judge the quality of a piece of research, it relies on full and comprehensible reporting of that research. For example, it is not possible to judge whether the research design fits the research question (Chapter 9) unless the question and design are both clearly stated in the report. Similarly, it is not possible to judge whether the research has enough cases to be convincing if the number of cases is not stated in the research report.

As you read more research papers, you will notice that the standard of research reporting in social science needs to improve in general, and we must play our part in reporting our research as simply as possible. Many of the key elements of empirical research studies are not clearly reported in so much social science. Much reporting is almost entirely incomprehensible, with long words and sentences, neologisms, uncensored lengthy extracts from software reports, and so on. Much of the remaining reporting is incomplete – presenting means without standard deviations, with no mention of missing data, or no description of how many respondents said one thing or another. In any report you read where it is not possible to find good information about the key elements, such as the research question(s), its scale and so on, the study should be treated as of no consequence, and can be ignored.

Otherwise, there are a number of factors to consider when judging the quality of a comprehensible and reasonably comprehensive research report. These could be considered in any order. However, it would make sense to start with the fit between the research question(s) and the study design (Chapter 9).

Other key factors to consider are the size of the study in terms of numbers of cases (Chapter 10), the amount of missing data (Chapter 12), and the quality of the data or measurements (Chapter 14). There are plenty of other considerations relevant to the internal validity of a study, including conflicts of interest, pre-specification of analyses, and appropriateness of analyses. However, here we will focus on the first four issues, because once these are decided then the other factors either generally fall into line as well, or at least cannot make our judgement worse.

Table 8.1 summarises these first four factors in four columns, with a generic description of research quality, judged in terms of each factor for each row, and it summarises each row with a security (padlock) rating in the final column. The overall rating suggests a research finding whose trustworthiness is at least at the level of the descriptions in that row. So 4🔒 suggests a study that is as secure as could reasonably be expected (because no research will be perfect), and 0🔒 represents a study that is so insecure that it adds nothing safe to our knowledge of social science. Of course, there are no objective criteria for deciding on any rating, or even on how many categories of ratings there should be. This is true of any such approach to judging quality (above). However, this table has been used by a large number of reviewers who have found it useful in sifting thousands of studies by quality, and they have found considerable agreement between themselves when allocating summary security ratings independently of each other.

The table's cell descriptions do not include numeric thresholds. This is deliberate and leaves control in the hands of the research reviewer. Anyway, things tend to settle down over the decisions in the four main columns. For example, the phrase a 'large number of cases' might be interpreted differently, depending upon the precise context, question or pay-off. There is also an interaction between the simple number of

**Table 8.1** A 'sieve' to assist in the estimation of trustworthiness

| Design | Scale | Missing data | Measurement quality | Rating |
|---|---|---|---|---|
| Strong design for research question | Large number of cases (per comparison group) | Minimal missing data, no impact on findings | Standardised, independent, reasonably accurate | 4 |
| Good design for research question | Medium number of cases (per comparison group) | Some missing data, possible impact on findings | Standardised, independent, some errors | 3 |
| Weak design for research question | Small number of cases (per comparison group) | Moderate missing data, likely impact on findings | Not standardised or independent, major possible errors | 2 |
| Very weak design for research question | Very small number of cases (per group) | High level of missing data, clear impact on findings | Weak measures, high level of error, or many outcomes | 1 |
| No consideration of design | A trivial scale of study | Huge amount of missing data, or not reported | Very weak measures | 0 |

cases, their completeness, representativeness of a wider set of cases, and the integrity of the way they have been allocated to groups. A 'large number of cases' would certainly be in the hundreds, but there is no precise figure such as 400 that can be set, other than as a rough guide. An excellent study might have one case below whatever threshold is suggested (399), and a weaker one might have one more (401).

Similarly, a true RCT might be considered a 'strong design for the research question' in one context, but there are other designs of equal or better ability to address other research questions. Some such designs may not even have been thought of yet (Chapter 9). An attrition rate of 2% might be crucial if the missing cases all had extreme scores in the same direction, whereas attrition of 10% might still yield reasonably secure results if there was an obvious reason for the dropout that was unbiased across groups and types of cases (Chapter 12). There is no clear threshold between 'minimal' attrition and worse that can be defended. It is based on what is known about the context and the nature of the missing cases, and where they appeared in the research process.

A suggested procedure for using the table would be to start with the first column, reading down the design descriptions until the research you are reading is at least as good as the descriptor in that row. Staying in the row achieved for the design, move to the next column and read down the descriptions, if needed, until the study is at least as good as the descriptor in that row. Then repeat this process for each column, moving down (never up) the rows, if needed, until the study is at least as good as the descriptor in that row. For any column, if it is not possible to discern the quality of the study from the available report(s) then the rating must be placed in the lowest (0) category. Each study sinks to its lowest level on these four key factors. The final column in the table gives the estimated security rating for that study (Gorard et al., 2017).

There is more complete guidance on design, scale, missing data and data quality in the rest of this part of the book, including practical exercises. Once you have read these further chapters you will be ready to make all of the judgements that Table 8.1 requires. If, and only if, you are happy with the security of any research finding then it makes sense to proceed to consideration of what the result might mean. The result might mean little or nothing of course. Security and substantive significance are not obviously related. Just as a weak result might appear exciting, so a secure result might be uninteresting or meaningless.

――― Exercise 8.1 ―――

Select a research paper or report in your own area of interest that involves in-depth data from interviews. Decide if the paper provides enough information on the strength or prevalence of any patterns or differences it reports. From what is reported, and using the approach described above, can you make your own tentative estimate of the strength or prevalence of any reported patterns or differences?

――― Exercise 8.2 ―――

Does the approach above help when judging the merits of a theoretical or conceptual paper as well as an empirical one? Discuss how to judge the rigour of a conceptual paper. Or is rigour not what we are looking for?

## JUDGING THE MEANING OF A FINDING

Chapters 4 and 5 illustrated several findings from both made-up and real datasets. But the chapters went no further than reporting the surface results. For example, there is no easy answer to how big a difference between two groups has to be for it to be taken seriously, whether for policy, practice or theory. In measuring the average heights of two groups of adults, a difference of a fraction of a millimetre can probably be ignored. Everyone can imagine that such a small difference, relative to the average height of an adult, could be the result of a tiny error in one measurement, or a minor variation caused by the method of sampling, or in the identification of the two groups, or by time of day for measuring each group (because we tend to be slightly taller when we wake up). On the other hand, even a difference this small could be of great importance in a physics experiment, perhaps because it is larger relative to the things being measured, or because the measurements were more precise, or the sample was millions of times larger.

The context and the meaning of the numbers involved in any result combine with estimates of measurement precision, the study's power of discrimination, the scale and variability of the measurement, and the sample size, to help a researcher make a judgement about whether a difference matters. So, on the one hand, there is no technical and easy way of saying that an observed difference in the data is important, or not. On the other hand, we all routinely make such sophisticated judgements all of the time, almost without noticing. We are naturally quite good at it.

Consider this analogy. You are in a supermarket (chain A) trying to buy an item – perhaps a number of tins of soup. You see the price, and are aware that another local supermarket (from chain B) is offering the same item for 4p less each. Which of the two supermarkets is selling this item for the lowest price? This sounds like a silly question. Supermarket B is offering the tin of soup for 4p less, and that is the lower price of the two. Yet researchers, both new and highly experienced, regularly seek help to decide on questions like this.

Perhaps these researchers have used an attitude scale and found that the male respondents had an average score of 3.7, and the females an average of 3.8. Clearly, the score for females is higher. But the researcher asks if it is 'really' bigger. And the answer is yes it is really bigger, as long as they measured, transcribed, recorded and analysed the individual scores correctly. The researcher usually, and wrongly, brushes these *caveats* aside as unimportant to them. What the researcher seems to want to ask is a combination of two very different questions. They want to know whether this apparent difference between the attitudes of men and women would also be true of some larger population of men and women. That is, they want to generalise their finding (see below). And they also want to know the really crucial thing, which is whether this difference of 0.1 is worth pursuing, substantively, practically or scientifically.

The supermarket analogy can help again here. It shows that the first question about generalisation is often not very important. If you want one tin, then it is not relevant whether other supermarkets in chain A generally charge 4p more than chain B for this soup. You know that this branch of supermarket A charges 4p more than supermarket B at this point of time for this transaction. This is all of the information you need to make your decision. And if the two chains have a standard pricing policy then you know that all branches will have the same price for this item, barring mistakes. Or if the tin is dented and appears on a shelf devoted to reduced cost items, then you know that the price will generally be higher elsewhere. In each of these situations and many others, the generalisation answer is trivial in the sense that it makes no practical difference to you. And even if you do want to consider the difference for entire chains of supermarkets, you are still then stuck with the second question about whether the difference is big enough to matter.

**Image 8.1** The cost of soup

The key issue is therefore the second one – is the saving of 4p sufficient to make you go to the other supermarket instead? The answer to this crucial question depends on such a lot of different factors that there is no definite answer for all people, and the answer cannot be a technical analytical one. The answer does not depend only on the simple (statistical) difference of 4p.

Yet each of us will come up with the right answer for ourselves quickly and easily, in practice. For example, if you wanted to buy a large number of tins of soup then the total saving could be large, and this might influence the decision in favour of going to the cheaper supermarket. If the other supermarket is a long way away, or you are short of time, this might influence the decision in favour of sticking with the more expensive supermarket. If your resources are large relative to a 4p saving, or 4p is minimal in relation to the cost of the tin, you may be more likely not to change supermarkets.

So it is perfectly proper that different people will come to different rational decisions, and there is no simple technical way of deciding. Yet, in reality, the decision is usually easy to make. In real life, we can almost instantly compare scale, quality, convenience, value for money and other factors, and then synthesise the results and decide what to do. The same is true with social science data analysis. It is really quite easy. What makes it only slightly harder than 'real life' is that as a researcher you must also explain fully to others, via a research report, why you decided to ignore or pursue the difference you found.

There are a number of factors to consider when examining a possible difference (or any other pattern or trend). These include the direction of the difference, the size of the difference relative to the scores themselves, the variability of the scores, their relative importance for subgroups of cases, the costs and benefits of getting the decision wrong in either direction, and the substantive and theoretical importance of the decision. That is a lot to take into account in one decision. The discussion so far has been about differences between groups or cases. But the same idea about judging results and explaining your judgement to others also applies to uncovering trends, or indeed any pattern in your data. And it is important to remember that a finding of a lack of pattern can be just as interesting as a pattern might be.

It is clear that, for any dataset, dividing the cases into any two or more subgroups will rarely yield exactly the same mean scores on all measures for both groups (Meehl, 1967). It is unlikely a priori that the people sitting on the left-hand side of a bus will have exactly the same average height as those sitting on the right. They are unlikely to report drinking exactly the same average number of cups of tea every day, and so on. A difference in scores or observations may, therefore, have no useful meaning at all.

Whether a difference is more than this, and is actually substantial and worthy of note, can depend on a number of factors. It depends on the size of the difference in relation to the scale in which the difference occurs (an observed difference of 2 feet may be important in comparing the heights of two people, but not in comparing flight distances between Europe and Australia). It also depends on the variability of all of the scores. It is harder to establish a clear difference between two sets of scores that have high levels of intrinsic variation, than between scores in which each member of each group produces roughly the same score as all other members of that group (hence the effect sizes in Chapter 5). The noteworthiness of a difference may also depend upon the benefits and dangers of missing a difference if it exists, or of assuming a difference if it does not exist (Cox, 2001).

All of these issues of scale, variability and cost are relevant even if the scores are measured precisely. But in reality, scores are seldom measured precisely, and social science measurements such as self-esteem, aspiration or occupational class will be subject to a very high level of measurement error. Measurement error is nearly always a bias in the scores (i.e. it is not random). People who do not respond to questions accurately (or at all) cannot be assumed to be similar to those who do. Children for whom a school has no prior attainment data, or no knowledge of their eligibility for free school meals, cannot be assumed to be the same as everyone else in other respects. A ruler that is too short and so overestimates heights will tend to do so again and again, uncompensated by any kind of random underestimates to match it. Even human (operator) error has been shown to be non-random, in such apparently neutral tasks as entering data into a computer.

So knowledge of the likely sources of error in any score, and an estimate of the range of measurement errors, constitute an additional and crucial part of deciding whether a difference between groups is big enough (to justify a substantive claim). The harder it is to measure something, the larger the errors in measurement will tend to be, and so the larger the difference would have to be, to be considered of substantive interest.

## GENERALITY

Having settled the security of any finding, it may then be interesting to consider to what extent the finding could also be true for any other set of cases – such as those in different times, places or institutions, or with different characteristics. This is referred to as **generalisability** or **generality**. It is only worth considering this issue if the finding is considered trustworthy enough to proceed. Whether findings generalise must be a matter of judgement (Gorard, 2006b). We do not know what the findings would have been for these wider cases (else we would not be seeking to generalise). We can only imagine what they could have been.

This judgement may be aided by considering how similar or different the other cases are to those in the actual study – in terms of values and variables that we know about. For example, if the research findings are based on a mixture of individuals with characteristics also found elsewhere then it is plausible to imagine that cases elsewhere with these same characteristics might be similar in terms of the substantive findings. The closer the match, the less variable the findings actually are between cases, and the more relevant the matching variables appear to be, the more likely the findings are to generalise.

Of course, if the findings of any study do not seem to generalise, this does not mean that the findings of the research are not valid, or useful and interesting in their own right. Whether this is so would be a completely separate judgement.

## THE WARRANT PRINCIPLE

Once the methodological and other challenges to the trustworthiness of a finding have been assessed, and the result is found to be sufficiently valid, you should look for its simplest explanation. This would be an explanation that requires making no further assumptions (or if not possible then the fewest additional assumptions). An easy way to judge whether any such conclusion drawn from the finding is **warranted** is to consider how else the evidence for it could be explained if the conclusion were not actually true (Gorard, 2013b). This warranting process is both creative and logical. Creative because the researcher should be trying to find simpler alternative explanations than the best explanation they have so far. And logical in then eliminating the less plausible and less parsimonious (i.e. more complicated) theories.

Imagine that we want to see if someone can influence the toss of a perfectly fair coin through their mental effort. This person might claim that if they visualise a result of heads, then the coin will land showing a head more often than a tail (or vice versa). However, because their mental influence is only a weak force, this increase in heads will only be by a small proportion.

We would need to devise a way of testing this claim. We would make this as fair as possible – by asking the person to pre-specify which of heads or tails they were going for. This would prevent their claiming a result either way after the event (the equivalent of dredging for success), because either heads or tails are likely to be at least slightly ahead. We could use our own set of standard coins selected from a larger set at random by observers, to prevent the possibility that the coin was rigged somehow. We would video the process and allow a panel of judges to confirm the result of each throw. We would insist on a decent size of sample (number of coin tosses), to reduce the possibility of a fluke result, and generally do anything we could to make the research as trustworthy as possible (the kinds of factors in Table 8.1). There is a considerable literature on strategies to overcome bias and confounds as far as possible, especially for experimental trials (Cook and Campbell, 1979).

Having done all of this, our simple finding after 1 million trials is that there were 51% heads and 49% tails. As far as we and other observers are concerned this was a fair test, and a million coin tosses is a lot. There is no missing data, and 100% agreement among the judges about the result of each throw. In terms of the factors in Table 8.1, this study and its finding are about as trustworthy as it is possible for research to be.

The point has been stressed to help all readers understand the importance of judging the trustworthiness of the study before even beginning to consider other explanations. If the person used their own coin, or was the only one to see the results and record them, or there were only 30 coin tosses, then we probably have no need for further explanation. The research would simply be declared not good enough to be worth worrying about. Note that this weaker kind of study, with a scale of 30 cases and so on, is an analogue of what much of current social science research looks like.

If we accept the finding for the present, does the 51% mean that the person's claim was correct? That they can mentally influence a coin toss at a distance? Not yet. Even if we have eliminated – or, more accurately, greatly reduced the chances of – a methodological explanation, there is still considerable work to be done before accepting 'telekinesis' as the explanation.

The result could be a fluke. In any such test of any size there is unlikely to be exactly the same number of heads and tails, even with unbiased results. So there is a 50% chance at the outset that any slight imbalance would be in favour of heads. The chances of its being a fluke decrease as the number of trials increases. In some research situations, such as coin tossing, we could calculate this decrease in likelihood quite precisely. In most research situations, however, the likelihood can only be an estimate. In all situations we can be certain of two things – that the chance explanation can never be discounted entirely, and that its likelihood is mostly a function of the scale of the research.

If not an error, and not a fluke, then perhaps we can think of an explanation other than telekinesis. This is where creativity comes in. Phrased in traditional terms, just because we do not accept either error or fluke as the most likely explanation, this does not mean that we should just accept any other single explanation. Maybe this is how a coin of this type always performs. Maybe it is to do with the position of the moon, or whatever.

Alternatively, the difference could be evidence that the claimant is correct. They really can influence the result. If the difference is judged a true effect, so that a person can mentally influence a coin toss, we should also consider the importance of this finding. This importance has at least two elements. The immediate practical outcome is probably negligible. If someone could guarantee odds of 3:1 in favour of heads on each toss then that would be different, and the difference over 1 million trials would be so great that there could be little doubt it was a true effect. On the other hand, even if the immediate practical importance is minor, if this were a true effect it would involve many changes in our theoretical understanding of important areas of physics and biology. This would be important knowledge for its own sake, and might also lead to more usable examples of mental influence at a distance. In fact, this revolution in thinking would be so great that many observers would conclude that 51% was not sufficient to accept it, even over 1 million trials. The finding makes so little immediate practical difference, but requires so much of an overhaul of existing knowledge, that it makes perfect sense to conclude that 51% is consistent with merely chance, bias or some other explanation.

This situation, of facing at least three kinds of explanation, is one faced by all researchers using whatever methods, once their data collection and analysis are complete. The finding could have no substantive significance at all (being due to chance or similar). It could be due to faults in the research (such as a selection effect in picking the coins). It could be a major discovery affecting our theoretical understanding of the world (that a person can influence events at a distance). Or it could be a combination of any of these. The proposed answer will be a matter of judgement. It should be an informed judgement, based on the best estimates of both chance and error, but it remains a judgement and so the researcher has to report or explain it (although many currently do not do so).

## WRITING ABOUT VALUES: A REAL EXAMPLE

Consider now a real-life example of the difficulty of deciding what a result means. This was an RCT involving 5,619 Key Stage 4 (KS4) students (aged 15 and 16) in England, individually randomised to an intervention or an active control (See et al., 2020). The two groups were balanced in terms of their prior attainment at school. Both groups received a sealed envelope, so that the trial was **blind**. The intervention involved students writing a brief essay in a registration period about what they valued about themselves. The control students were asked to write a brief essay about what they valued in others. The process was assessed as costing schools less than £2 to implement.

The idea, which had already shown some promise elsewhere, was that thinking about personal worth just before a key examination would boost results for the students known to be disadvantaged (those eligible for free school meals due to family poverty). The study was independently evaluated as having a high trustworthiness rating in terms of appropriate design, large scale, low attrition, and quality of outcome measured (official KS4 results). It showed that the disadvantaged students writing about themselves had better subsequent examination results than those in the control, with an effect size of +0.05 (see Chapter 5).

Assuming that we trust the result, what does it mean? Traditionally, 0.05 is not considered a substantial effect size, and is too close to zero for us to be completely confident in, despite the quality of the research behind it. Nevertheless, this is such an easy, cheap and non-intrusive intervention that it might be worth schools using it more routinely. They have nothing to lose. It may help disadvantaged students and so reduce the poverty attainment gap, or it may make no difference. It does not seem to do any harm. All of this is so, even though an effect size of 0.05 in a different context is perhaps not worth pursuing.

## CONCLUSION

The procedure for judging the quality of research described here is intended to be as inclusive as possible. It is deliberately non-specific about the kind of data involved in any study, because the kind of data collected is independent of issues such as design, scale and attrition.

Having established a trustworthy result, there are no simple rules we can apply to decide whether an apparent difference, pattern, change or trend in any dataset is actually worth pursuing. We could posit a rule of thumb which is that we need to be sure that any effect sizes we continue to work with are substantial enough to be worth it (Cox, 2001). Clearly, this judgement depends on the variability of the phenomenon, its scale, and its relative costs and benefits, as illustrated in the examples above. It also depends on the acknowledged ratio of the effect to potential error. Therefore, a difference that is worth working with will usually be clear and obvious from a fairly simple inspection of the data. If we have to dredge deeply for any 'effect', then it is probably pathological to believe that anything useful will come out of it. We cannot specify the minimum size needed for an effect, nor can we use standardised tables of the meanings of effect sizes.

## Notes on selected exercises

### Exercise 8.1

It is alarming that many papers and research reports do not provide much detail about the data involved. This is perhaps especially the case for work involving in-depth data where no numbers are attached to the claims. Notice that this does not prevent the authors still claiming the existence of things like patterns and differences. It just means that the readers can have no real idea of the prevalence of any pattern or trend. Instead of explicit numbers, vague numeric terms like 'many', 'some' and 'most' are used. The dataset is seldom available. And the data examples quoted are usually unclear in origin. Are they average examples, ideal ones, or the best expressed or the most extreme? Usually, from what is reported by many other researchers, I have no way of making my own estimate of the strength or prevalence of any reported patterns or differences. Surely this has to change if we demand it, or at least if we just ignore any research that does not provide this very basic information. A lot of this chapter has been about numeric data. It is important to recall that the difficulties and problems are almost certainly worse, if somewhat harder to demonstrate, with non-numeric data.

### Exercise 8.2

Table 8.1 assumes that the judgement of quality is about empirical work. Judgements about more theoretical work would be more concerned with whether the ideas were logical, productive and useful. Of course, it is just as important that such a paper is clearly and simply argued.

## Further exercises to try

Find a research article using numeric data, one that you want to read anyway. Using the approach from Table 8.1, assess how trustworthy you think the study is. Discuss the difficulties of making such a judgement. Ignoring how trustworthy the research is, if you assume that the findings are correct, is their interpretation by the researcher warranted? Do the reported implications follow logically from the findings or is there an obvious simpler explanation?

## Suggestions for further reading

A relatively easy-to-read book on the importance of interpreting your numeric data, and creating a logical narrative to describe this process to your readers:
Abelson. R. (1995) *Statistics as Principled Argument*. Hove: Psychology Press.

A book on understanding uncertainty and probability:
Gigerenzer, G. (2002) *Reckoning with Risk*. London: Penguin.

*(Continued)*

A brief chapter explaining more about the idea of warrants for research claims:
Chapter 4 in Gorard, S. (2013b) *Research Design: Robust Approaches for the Social Sciences*. London: Sage.

Read about the James Lind Initiative to promote critical thinking when assessing research claims. There are many easy-to-understand resources here, even though they have a health focus:
Chalmers, I., Atkinson, P., Badenoch, D., Glasziou, P., Austvoll-Dahlgreen, A., Oxman, A. and Clarke, M. (2019) The James Lind Initiative: Books, websites and databases to promote critical thinking about treatment claims, 2003 to 2018, *Research Involvement and Engagement*, 5, 6. https://doi.org/10.1186/s40900-019-0138-2

# 9
# RESEARCH DESIGNS

## SUMMARY

As shown in Chapter 8, the quality of a research design is a factor to consider when judging the trustworthiness of research. Designing a research study is a relatively simple process that pays immediate dividends in terms of research quality and ease of reporting. Research design links the research questions to the procedures of the research study, making the analysis easier and the conclusions stronger. Given these major advantages, it is surprising that so few methods resources and courses emphasise research design as much as they should, and that so few researchers mention their designs in their reports and papers. This chapter describes what a research design is, the elements that make up a design, a number of common designs in use in social science, and the (lack of) relationship between research design and research methods.

## RESEARCH QUESTIONS AND DESIGNS

All good research studies will have an underlying design even if this is not reported, just as a well-built house has a foundation and an architectural plan even if these are not visible to people living in the house. The research design should stem from the research question(s) to be addressed by the study, and should suit the type of question being asked. Conducting research without considering its design is as ill-advised as trying to build a house without foundations.

Research should start with at least informal research questions, from which all else follows. These questions should be as precise and brief as possible. More short questions are better than fewer complicated or double-barrelled ones. These questions then lead naturally to the design of your research (described below). If, for example, you want to compare two groups then you need evidence from two groups, whatever data collection or analysis method you then decide to use. Yet, when you read the research literature in your field you will soon realise that the majority of studies have no clear research design, or use an inappropriate one.

'Research design' refers to the structure and organisation of a research project. In particular, it refers to the cases (participants) involved in any study, the ways in which those cases can be allocated to subgroups for comparison, the timing and sequence of any data collection episodes, and any interventions of interest to the researcher that could affect the research outcomes (de Vaus, 2001; Shadish et al., 2002). These are the main elements of a design (Gorard, 2013b). For example, the elements of timing and comparison groups for eight common design terms are summarised in Table 9.1, and then explained further below.

**Table 9.1** Four common research designs

|  | No comparator | Fair comparator |
|---|---|---|
| One-off data collection | e.g. case study | e.g. cross-sectional, or comparative |
| Repeated data collection | e.g. longitudinal, time series, or trend | e.g. experimental, or regression discontinuity |

When data is collected repeatedly from one group of cases the design is often called **longitudinal**, which emphasises the time element (e.g. Gorard, 2015b). When data is collected from two or more groups on one occasion the design can be called **cross-sectional** or 'comparative', emphasising that the cases are in subgroups (e.g. See et al., 2015). When there are two or more groups, and some groups receive an intervention that other groups do not, then this forms the basis for a design that could be called **experimental** or **quasi-experimental** (e.g. Siddiqui et al., 2015). As their use of different elements and their names suggest, these designs have different roles. Experimental-type studies are good at answering causal research questions, such as whether the intervention made a difference. The other designs can really only address descriptive questions. Cross-sectional studies are good at answering comparative questions, such as whether one group of people has a higher income than another (see Chapter 5). Longitudinal studies are useful for identifying risk factors in biographical data (see Chapter 19).

## THE ELEMENTS OF DESIGN

The **'cases'** in a study are the individuals, organisations or objects selected to take part in the research. All possible cases that could take part form the **population** for the study (Chapter 10). All other things being equal, the more cases there are in any study, the more trustworthy its findings are. If possible, all eligible cases should be used since this will minimise bias (as opposed to selecting only some cases from the population). If this is not possible, then a subset of cases can be selected from the population. If these cases are selected randomly this produces the least bias (other than using all cases). If any other form of selection is used, such as convenience sampling, the research is already considerably weakened in terms of its generality (Chapter 8).

The planned scale of research is important, but recording and reporting how many cases are missing (non-response, dropout, non-contactable) or have key missing data (Chapter 12) is at least as important. All research reports should include a minimum set of information about the cases, such as the population, how many cases were selected, how many did not take part, and how many were later excluded for any other reason. In general, and all other things being equal, the fewer missing cases there are in any study, the more trustworthy its findings are (Chapter 8).

A simple research question like 'Do girls do better than boys at school?' is comparative ('cross-sectional'). It is a more interesting question than could be addressed by a case study (i.e. a study of a case without any comparison, duration or intervention). If the research question is comparative, this means that the researcher wants to make a comparison, as here between boys and girls. It is better (more convincing) to specify this comparison in advance in the research questions, rather than dredging the data after its collection looking for possible differences between any number of subgroups. This pre-specification would include how the comparator groups were created, and which variables or characteristics the groups will be compared on. Where the comparator groups occur naturally (boys versus girls perhaps), they can be useful for making descriptive comparative claims using a cross-sectional design (such as that girls get better exam results than boys).

A question like 'Is the attainment gap between boys and girls at school growing over time for each cohort?' would also be comparative, but is about change over time more than differences between groups. An appropriate design could be called a 'trend study', comparing successive cohorts of boys and girls, and it would answer a non-causal question. Of course other designs can be created that would also be appropriate, but a repeated cross-sectional design would be a good approach here. This would be like two separate comparative studies but joined together – one earlier in time and one later.

If instead the question was 'Does the attainment gap between boys and girls at school increase as they get older?', then this is again comparative, about change and non-causal. But it is longitudinal or biographical, not about repeated cross-sections. Therefore, a panel or cohort (longitudinal) design would be appropriate. This would again have to involve earlier and later measurements, but for the same cases on both occasions.

Trend, or repeated cross-sectional, designs are useful for describing historical changes over time, but of limited use in explanatory research. The health of a nation may be improving (over successive generations), even if the health of each individual in that nation is declining (as they age). Longitudinal, or panel or cohort, designs are useful for tracking changes over the life course of individuals (or cases), especially when looking for possible risk factors for subsequent events. The link between smoking and lung cancer was first discovered via this approach. However, neither design has been shown to be much good at predicting future events. Both are also known to be misleading when addressing causal questions. This is partly due to attrition of cases as time goes by, and other changes arising in society that could not have been known at the outset, but which could influence the outcomes.

**Image 9.1  Predicting the future?**

Adding an intervention or treatment to a longitudinal or other study does not make it a useful test of causation in itself. Before-and-after designs like this can be very misleading if they are treated as portraying a causal model. It is well known that correlation or mere association between two variables is not good evidence of causation. But nor are before-and-after changes. Giving school students an extra holiday and then discovering that their reading ability is improved a year later is not evidence that the holiday caused the improvement in reading. The design needs a **counterfactual** as well. What would have happened to these students if they had not had the holiday? Probably most students would improve over a year, regardless of other factors. Comparing those students with the extra holiday with another equivalent group which did not receive the extra holiday is crucial. The two groups must be as similar as possible to begin with – hence the randomised control trial (RCT) or experimental design.

It is hard to make some kinds of research claim using **heterogeneous groups**, such as boys and girls, rich and poor, employed and unemployed, and so on. So researchers often want to create equivalent groups at the outset, such as the **treatment** and **control groups** in an experimental design. For example, the research question 'Is the gap in attainment between boys and girls caused by the nature of the assessments used?' is a causal one. Addressing this might involve a **natural experiment** or RCT. The surest way to approach this is to randomly allocate each case individually (whether a boy or a girl) to one of the subgroups in the study – where each group has different kinds of assessment. This is the safest way, because it caters for differences in terms of unknown variables that can never be matched, as well as known variables that could have been matched using a weaker design.

If randomisation is not possible then we could use some kind of matching of cases in terms of their known characteristics. However, even when it is successful such a procedure only matches the two groups in gross terms, and based on only those variables that are available. For example, matching the average age of participants in two groups does not mean that the groups are similar or unbiased in terms of motivation. Randomisation, on the other hand, deals with all possible differences at the same time. This is what makes randomisation very powerful.

An experiment with two groups allocated at random, and tested after one group has had a different intervention than the other group, is a simple design. Adding a test for both groups before the intervention as well allows researchers to compare the gain scores (from pre- to post-intervention) between groups. This design allows for slight variation in the initial scores between groups, because the groups are likely not to be perfectly balanced at the pre-intervention test (that is part of what 'random' means). Generally, experimental trials are powerful, relatively inexpensive, easy to conduct, and they make subsequent analysis simpler. You just have to compare the results for two groups (as in the example in Chapter 5). That is all the analysis you will need. If an RCT has been designed properly, you can then begin to make warranted *causal* claims about your findings.

A simple example of an RCT might involve testing the efficacy of a new lecture plan for teaching a particular aspect of mathematics. A large sample is randomly divided into two groups. Both groups sit a test of their understanding of the mathematical concept, giving the researcher a pre-intervention test score. One group is then given lectures on the relevant topic in the usual way. This is the control group. Another group is given lectures using the new lecture plan. This is the experimental treatment group. Subsequently, both groups sit a further test of their understanding of the mathematical concept, giving the researcher a post-intervention score. The difference between the pre- and post-intervention scores yields a gain score for each student. If there is little or no difference between the average gain scores for each group then we can say that the intervention has been ineffective (it is no better than the normal lectures). If the average gain score is higher for the treatment group then this is evidence that the treatment has been beneficial. If the gain score is higher for the control group this is evidence that the intervention has been harmful. We would probably convert the difference between groups into an effect size (Chapter 5)

We would next assess the trustworthiness of the study (as in Chapter 8). If we are happy that the study is strong, then the next stage would be to assess the size of the difference, at least partly in relation to the cost of the intervention. Then we could make recommendations about whether the new approach to teaching should be adopted in practice or not.

Examples of analyses using each of the designs discussed here appear throughout the book. However, the elements of a design can be combined in many more ways than described so far, and only a few of these ways will have clearly recognisable names like 'longitudinal'. Be creative, and come up with your own designs to suit your own research questions, using the elements as building bricks (see more ideas in Gorard, 2013b).

---

### Exercise 9.1

Find an empirical research article in your own area of interest that makes a comparative claim.

a   Does it have an explicit comparator?
b   Does it explain how many other comparisons were made as part of this study?
c   If a lot of other comparisons were made but not reported, would this influence your judgement about the trustworthiness of the comparison reports?
d   Were the other comparisons specified before the study was conducted? Was the result of the comparison predicted before the study was conducted? What difference would this make to your judgement of the findings?

---

### Exercise 9.2

Imagine that you read an article in a social science journal showing how a small early investment in overcoming personal poverty leads to significant impact on the chances of poverty in later life. The researchers have evaluated a programme to provide pocket money to children and young adults from the poorest 5% of families in one area. By age 16, the families of well over a third of these young people are no longer among the poorest 5% in that area. In fact, a few families are now among the richest 50%. This is an impressive result for what amounts to a very small investment, and in their paper the researchers call for their programme to be rolled out across the country.

a   What is the key logical error being made by the researchers here?
b   Using the warrant principle, if their conclusion was actually incorrect then how else can you explain their dramatic findings?

---

## THE INDEPENDENCE OF RESEARCH DESIGNS

It is important to recognise that none of the elements of a research design referred to are necessarily related to particular methods of data collection or analysis. For example, if a certain number of cases are needed to make a convincing research claim then this is true however that data was collected. If a particular sample size is needed to make a believable finding about what people state about some phenomenon, it does not matter whether those people make the statement to the researcher via a recording, a survey form, a computer or conversationally in passing. All other things being equal, the larger the sample, the stronger and more convincing the research results will be. And all other things being equal, a random sample will provide a less biased estimate of a more general population than any other kind of sample. Research craft knowledge like this remains valid whatever kind of data is then collected from that sample.

A study that followed infants from birth to adolescence, weighing them in January every year, would be longitudinal in design. A study that followed infants from birth to adolescence, interviewing their parents about the child's happiness every year, would also be longitudinal. A study that did both of these would still be longitudinal, even though some commentators would pointlessly categorise the first study as quantitative, the second as qualitative, and the third as 'mixed methods'. In each example the design – longitudinal, or collecting data from the same cases repeatedly over a period of time – is the same. This illustrates that the design of a study does not entail a specific form of data to be collected. Nor does it entail any specific method of analysis. Nor is any method tied to a specific research design.

The implication for handling numbers and the analyses in later chapters is that, while your analyses must take the research design into account, there is no difference in analysing a set of data which is dependent upon whether it was gathered face-to-face, using information technology, or on paper. Each approach may require some special skills, but its overall logic remains the same.

## CONCLUSION

There are many more varieties of design than those mentioned above. There can be any number of groups, episodes and types of data collection, interventions, methods of allocating cases to groups, and so on. These can be assembled in any order to create new and innovative designs. There is no one good design because the suitability of a design depends so much on the research questions to be answered. A researcher using a case study to address a causal question is making a clear mistake. A researcher using a randomised control trial to address a simple comparative question is making an equally clear mistake. There is no 'gold standard' design. Designs are not to be selected on the basis of purported research paradigms. Nor do they have anything to do with the methods of data collection and analysis used.

Design is chiefly about care and attention to detail, motivated by a passion for the safety of our research-based conclusions (see Chapter 8). Preparation before conducting research, and making allowance at the start for the kinds of claims you will want to make afterwards, will generate better research. It really will. It will also generate results that are easier to analyse. Complex statistical methods cannot be used post hoc to overcome design problems or deficiencies in datasets. Rather, a good design means you only need to use simple statistics (Wright, 2003). Good design will help create research conclusions that are firmly warranted. These will also be easier to communicate to a wide audience.

Of course a good design can fail because of poor or inappropriate data. But a poor design remains poor however good your data is, and whatever type it is. A poor design, including no design at all, means that you will generally not be able to draw the kinds of conclusions you want to. It means that your time, and the time of everyone involved in or reading about the research, will have been largely wasted.

At its simplest, research design is about convincing a wider audience of sceptical people that the conclusions of the research, perhaps underlying important real-life decisions, are as safe as possible (Chapter 8). If we are going to risk the happiness of a family by removing a child from their parents, risk public safety by releasing prisoners early, or spend public money on almost any societal intervention, then we should want the research underlying that decision to be as strong as possible. Such policy decisions might be correct, or they might be a wasted opportunity or worse. It is the task of social scientists to help make such decisions as foolproof as possible. At present, despite a small amount of excellent work in every field, this is just not happening sufficiently. Maybe you can help to change that in the future. This book should help you.

## Notes on selected exercises

### Exercise 9.1

a   It is shocking how easy it is to find simple flaws like implied comparative claims without a comparator.
b   The number of comparisons is important, because it provides the "denominator" for how convincing any claim is.
c   Quite how remarkable a pattern in your interview or survey data is depends partly on how many patterns you sought, and whether you predicted the pattern before collecting the data. I seldom see reports that make this explicit, which is shocking once you realise how important this factor is in being convinced by any findings.
d   Quite clearly, if the comparisons were built into the research design from the start then they ought to be more convincing to the reader. I wonder, though, whether the readers and critical consumers of social science remember to make this distinction when reading research.

### Exercise 9.2

a   The key logical error in this claim is known as post hoc ergo propter hoc, which literally means after this so because of this. We have no idea what would have happened if the pocket money intervention had not taken place. More formally, we could say that the result has no counterfactual. This would be a very unsafe result to use as the basis for rolling out the intervention more widely, assuming that this was the only evidence available and the only reason for doing so.

b   If the giving of pocket money was not effective, then plenty of other reasons are available to explain the lifting of so many families out of poverty. One is regression to the mean. The poorest 5% have less room to become poorer over time, but plenty of room to become less poor. Those in the middle of the income distribution can become richer or poorer. Thus, it is perfectly possible that this change would have happened over time anyway. Without a comparator group not receiving the pocket money, we do not know whether the intervention was effective or not. Hence the need for RCTs and similar, so that we can provide the best evidence to reduce poverty.

## Further exercises to try

Find a research paper that uses numeric data. Try to identify the research question(s). Try to identify the research design. If possible, judge how appropriate the design is to address the research question(s).

## Suggestions for further reading

A chapter on identifying appropriate questions to research:
Chapter 3 in Gorard, S. (2013b) *Research Design: Robust Approaches for the Social Sciences*. London: Sage.

Read the first few introductory chapters, for a useful overview of design:
de Vaus, D. (2001) *Research Design in Social Research*. London: Sage.

Thoughts on identifying causal models in social science:
Arjas, E. (2001) Causal analysis and statistics: A social sciences perspective, *European Sociological Review*, 17(1), 59-64.

A relatively complex book on the rarer designs, sometimes used in psychology:
Edwards, A. (1972) *Experimental Design in Psychological Research*. New York: Holt, Rinehart and Winston.

This is a book about research methods rather than design (see this chapter) that might help someone inexperienced who wanted to run a small empirical study for themselves:
Babbie, E. (2016) *The Practice of Social Research*. Wadsworth, CA: Cengage.

# 10
## SAMPLING AND POPULATIONS

### SUMMARY

This chapter looks at the cases involved in any analysis, whether from your own study or one that appears in an article you are reading. The cases form part of the research design in any study. They are the units from which data has been gathered, and the units for the analysis that follows. Cases are often individual people in social science research, but they could also be organisations such as hospitals, objects such as paintings or documents, or indeed almost anything. Cases are selected from the population of interest. The chapter looks at working with population data, selecting a sample, and how to judge the quality of samples. The quality and scale of the cases in any study are an important element when judging the trustworthiness of a finding.

### AUTHORITY AND GENERALITY

Firefighting could be considered a dangerous job. But suppose that you discovered that the number of deaths per person among professional firefighters in one city was actually lower than the number of deaths per person for everyone else in the city. Perhaps in one year the figures were 10 deaths per thousand firefighters over a period of time, and 20 deaths per thousand for everyone else. Does this mean that just living in a city is somehow more dangerous than being a firefighter living in that city? Does it mean that becoming a firefighter bestows some kind of protection from the dangers of living in the city?

───────── **Exercise 10.1** ─────────
What are the other possible, or more plausible, explanations for this difference? Discuss.

These questions are linked to the difference between the generality and security of a research finding. The security (or trustworthiness, authority or internal validity) of a research finding refers to how convincing the finding is. Or how persuasive any finding would be to a sceptical audience (Chapter 8). It concerns a number of related factors such as the design used in the research (Chapter 8), and how high quality the collected data is (Chapter 14). However, another major factor is the quality of the sample, or cases used to collect the data from. This quality is partly about the scale of the sample. In general, the larger a sample is, the more authority it has. Readers will correctly be much more convinced by the findings from a survey of 800 people than from a survey of 8 or even 80 (all other things being equal). But almost as important is completeness of the responses. Missing data, cases refusing to take part, and dropout are all causes of bias (Chapter 12).

The generality (or generalisability, or external validity) of a research finding, on the other hand, is about whether the finding is also applicable to situations, processes or cases that were not involved in the research. As should be obvious, this issue is very much secondary to the issue of authority. It only makes sense to worry about the generality of any research finding if it is first judged to be trustworthy. Put another way, if we do not trust the findings of a piece of research with its specified cases, then it does not matter at all whether those untrustworthy findings would also be true of other cases not used in the study, Other methods resources about the use of numbers tend to give this issue of generality much more importance than it deserves.

For the sample or cases in any study, the issue of generality is whether findings based on the sample are likely to be more generally true of the population from which the sample was selected, or perhaps of other samples that could have been selected for the study. This is again partly about the scale and also about lack of bias in the study, but it is more directly about whether the overall pattern (or characteristics) of the sample would match the pattern for the known population.

In the firefighting example, we would first have to decide whether we accepted that the finding of 10 deaths per thousand for firefighters was accurate and to be trusted. Only then does it make sense to consider whether the finding would or should also be true of the wider population of the city. In this example, even if we completely trust the 10 per thousand figure it is not clear that we would expect the same figure to be more generally true across the city. Firefighters might tend to be younger than the average for all residents, and might also be subject to basic health and fitness checks as part of their selection for the job. Both factors would tend to reduce the death rate of firefighters while in service, even if the job were intrinsically dangerous. A fairer comparison would be between firefighters and other city residents of the same age and initial fitness. Authority and generality are different aspects of research, and any study can have authority without generality. A study might also have generality without authority, but this is irrelevant.

## POPULATIONS AND SAMPLES

It is easy to see that the simplest way to convince a sceptical audience that the cases in the research genuinely represent the entire population of interest is to use the entire population as cases. So, if it is possible on resource, practical and ethical grounds to conduct the research using all possible cases of interest, then this is the safest scientific way forward.

A dictionary might define a **sample** as 'a small part, or subset of examples, intended to show what a larger whole or complete set is like', and this is a reasonable version of what a sample is in research. The 'whole' is usually referred to as the population, and so a sample is a smaller set of cases (things) that is intended to represent the population of all cases as accurately as possible. Looked at the other way around, the population for any research is a set of all of the cases that *could* have been in the sample chosen for the research.

---
### Exercise 10.2
---

A researcher is interested in studying one hospital, and arranges interviews with 20 nurses who work in that hospital. The 20 nurses form the sample for this study. What is the population for this study?

What would the population be if:

a   There were 30 nurses in the sample?
b   The researcher conducted a survey of the nurses instead of doing interviews?
c   The researcher also interviewed a few patients?

---

The cases in any research are the individuals, organisations or objects selected to take part. The cases can form a population, in the sense that the study involved all eligible relevant cases. This is a kind of **census**. Or a sample of cases can be selected from a population. Thus, a sample is a selection or subset of all eligible relevant cases for any research study – the latter being the population. The population for any sample of any kind is all of the cases known to have a non-zero chance of being in the sample. If the population is unknown then there is no population (in this technical sense). This in turn means that there can be no sample (in that same sense).

Picking a street and knocking on every door in that street in order to collect data from residents would be to work with a population (i.e. the residents of the whole street). The population would be all residents at home in that street when you called (whether they answered the door or not). Picking a street and then picking a random subset of households on that street, and then knocking only on those doors in that street in order to collect data, would be to work with a sample from the population.

Stopping people in the street as they pass, in order to collect data, is not really working with a sample because it is not clear what any achieved sample would be a sample of. The people may or may not be from that street. The population in theory is all of the people who walked along that street while you were collecting data. But these cases are not known; even their total number is unknown. In this example, the cases more closely represent just a group of people stopped on a street than either a sample or a population. This is not a problem in itself, and high-quality data might still be obtained from such passers-by. The problem only arises if the researcher tries to claim that these cases are representative of some larger population, and/or does so using analytical techniques that would only be valid for a genuine sample.

Examples of populations include:

- All hospitals in one authority
- The national unemployment figures
- All pupils in one school.

Examples of population studies also include the 1958 National Child Development Study and the 1970 British Cohort Study which are following all of the babies born in Great Britain in one week (whose families agreed to take part). Of course, some families did not agree to take part, and some cases have dropped out since the start, but these factors simply make the population in both studies incomplete. They do not make either cohort into a sample study. Further examples of population studies would include the national census of population, the set of cases being randomly allocated to treatment groups in an experimental design, or a survey of all of the prisoners in one prison.

A population, in this sense, can be of people or of other types of cases such as institutions or books. What all of these population examples have in common is that the study involves or attempts to involve every relevant and eligible case. If a study surveys all of the prisoners in one prison this is the population for that study. There can be no prisoners in that prison that are not meant to be part of the study, while prisoners in other prisons and people not in any prison had no chance of being in the study. The latter are not, therefore, part of the population for the study.

A sample, as a subset of the population, is already incomplete by definition. And the very process of sampling can introduce bias over and above any problems that might arise with population data (such as unknown missing cases, non-responders, missing data from existing cases, and erroneous data). Regardless of how a sample is selected, none of the problems of bias can be solved technically, but all must be taken into account during analysis. This means that most real-life samples should be treated in the same way as incomplete population data. In practice, they represent a partial set of the eligible data. Their authority and the substantive importance of findings based on the sample must be treated separately from the issue of generality. If relevant, generality has to be established subsequently by direct comparison between the sample and any further cases to which it is being generalised.

## WORKING WITH POPULATIONS

Choosing to work with a population is a research design issue, and so is independent of the methods of data collection (we might interview the cases or measure something about them, or both, for example). In a lot of social science, one of the aims of research is generalisation to the population. The beauty of working with population data is that this generalisation is already achieved, by definition. No further analysis is needed, or appropriate, in order to generalise. This makes population research intrinsically more rigorous, and more convincing in its claims, than equivalent studies involving samples.

Not having to worry about generality also allows analysts to have a greater focus on the things that really matter – such as the meaning of the data, its quality and completeness. A researcher may still wish to generalise from the population in the study to other populations, but this can only be a judgement-based generalisation. Such generalisation is done on a case-by-case basis, treating the research population as a new form of case. For example, it may be that the results of a survey in one hospital provide lessons for other hospitals, even in other countries, and perhaps even for other public institutions like schools and prisons. But no statistical generalisation is possible or needed to provide the basis for those lessons.

If you want to analyse the 16-year-old school examination results of all students in England for one year group then the population will be around 700,000 students. Because the data on the results for all 700,000 already exists it is as easy to analyse the existing results for all students as it would be for a small sample. There is no excuse for using a sample. The number of cases required would be 700,000, or as many of these as have appropriate data. It takes no longer to download all cases than to ask for some of them. It takes no longer to analyse 700,000 cases than 700. The summary tables (e.g. of the mean of a variable) will be the same size for 700,000 as for 700, and the research report will take the same time to write. The scale of most research in itself is not a major factor in the time or cost involved. It is often not clear why samples are used at all, or why, if they are used, they are frequently so small.

It is perfectly proper to look at patterns within population data, or differences between subgroups. Dividing a population into heterogeneous subgroups generally produces groups that are themselves mini-populations, and all of the advantages and restrictions outlined above still apply. For example, if the study involves all of the patients in one hospital, then dividing the patients into two groups by their age (e.g. above 30 or not) produces two further populations – all of the patients above 30, and all of the others, in one hospital. Claims about the comparisons, differences, trends or patterns in these subgroups are still claims about populations. No analysis based on sampling theory is involved or can be reported. Whether the findings for either group are also likely to be true for the population is a redundant analytical question. The findings *are* for the population.

Analysis of populations is therefore as simple as it is possible to be, and can involve totals, means, percentages, graphs, correlations, indices of inequality and so on, just as with any numeric data. Population data can also be safely modelled using techniques such as regression analysis, as long as care is taken that the software involved is not making default decisions about the model on the basis of things like covert significance tests (Chapter 6).

There are many other advantages to working with populations. The first and most important of these is the strength of the study. As shown in Chapter 8, the absolute scale of any study is a key element in judging how trustworthy it is. But using a population for your study means much more than just the possibility of a large number of cases. It also means no potential cases are left out and so there is no chance of sampling bias. Using a population is the strongest approach possible.

Of course, it is unlikely that any real dataset, even for a population, will actually be complete. The UK census of households every 10 years misses some residents, such as those away from home for a long period, the homeless, and a minority who cannot or will not complete the form. This does not make the UK population census into any kind of sample. It is merely an incomplete census, as *all* population data will be in real life. Therefore, the key issue for analysis does not concern generalisation but consideration of the missing cases and data, and how these might influence any findings (Chapter 12).

## THE KINDS OF SAMPLE

The reason for not using the population, and focusing research efforts on a subset of cases or sample, must be clear in order for a research design to be convincing to a sceptical audience. The first, and perhaps most commonly omitted, stage in sampling is deciding why you need to use a sample at all. Not all research is based on samples. Even discounting solely theoretical writing, as shown above it is not clear that all empirical research should involve a sample.

It is only your previously defined population that any sample should be drawn from, and which the sample will be representative of. Therefore if, for example, you only have the resources to carry out research in the immediate area of your home or institution, you cannot have a national or regional population. Anything you discover in your research will apply only to your immediate area. A sample drawn from the nurses in one health authority has nurses in that authority as its population. This may sound obvious, but is easily overlooked in practice. Other health authorities may note your findings with interest but logically there is no contradiction if they deny that the results are relevant to them (because many of the local socio-economic conditions or budgets differ between authority areas perhaps). The researcher can only set out to generalise in this strict sense to the population from which the sample was drawn.

Samples can vary in terms of how cases are selected, and the common types of sampling include random, stratified random, cluster random and non-probability. These are discussed in a kind of descending order of desirability.

## Random selection

In an ideal study, and if you have to sample, you will be selecting cases from the population at random (by chance) to form your sample. Thus, you need to start with a list of all cases in the population and give each of them a non-zero chance of being selected.

**Image 10.1**  Randomisation

A random number generator (computer software, random number table, dice, or a hat with lottery tickets in it perhaps) should be used to select cases one after another from the population. This means, of course, that the sample could be very strange and unrepresentative of the population. However, the probability of this is small (by definition, since extreme distributions are less likely than representative ones). The larger the sample is, the less likely such a 'freak' selection is. Chapter 11 discusses in more detail what randomness is, and Chapter 3 describes some easy methods for generating random numbers.

**Random sampling** is free of the systematic bias that might stem from choices made by the researcher. Apart from the practical problems of obtaining an accurate population list to work from, which is common to most approaches, the technique of random sampling is also the easiest method available.

## Stratified sampling

In a random sample the distribution of the characteristics of cases will be left to chance. In selecting 1,000 people from a general population, all could be male or all could be aged 13. This is very unlikely, and is more unlikely when the sample size is large. One can imagine rolling 4 sixes in succession with a die, but not 1,000. Where there is special concern over the small size of the sample, or the low frequency of one or more population characteristics, an alternative approach is to use **stratified random sampling**. Here, cases are selected in proportion to one or more characteristics in the population. For example, if employment is considered relevant to the study and the population is 58% employed, then the sample can be constrained so that it must be 58% employed. In effect, the researcher creates two populations, one of unemployed from whom 42% of the eventual cases are selected, and one of employed from whom 58% of cases are selected. Selection within these sub-populations can and should still be random if possible.

The number and type of characteristics used in this way (known as strata) are chosen by the researcher on theoretical grounds of relevance to the study. The researcher must use expert knowledge to decide which characteristics of the population could be relevant for the study findings, and then work out the pattern of distribution of these characteristics. This is not always an easy task, as the characteristics may need to be considered in interaction, and the researcher may need to carry out a census anyway to uncover the characteristics of the population. So this approach to sampling is considerably harder work than simple random selection. The researcher needs a good reason not to use random sampling.

The stratified approach can lead to a high-quality sample by reducing the risk of a 'freaky' result, at least in terms of the strata characteristics. Its problems include the fact that it can require decisions about complex categories (race, occupation) or on sensitive issues (income, age). Also, if several background characteristics are used then the selection process becomes difficult as each variable interacts with the other. If both sex and occupational class are used, then not only must the overall proportions for sex and class be correct, but so must the proportions for sex within each class (e.g. if 23% of the population are female and professional, this must be reflected in the sample). If you also, quite reasonably, considered ethnicity or area of residence as important factors then the calculations quickly become mind-boggling. You would need to know the proportion of the population who were white, male, professional background, and living in each area. Then you would need to reflect this in your sample. And you would need a large sample in order not to have increasingly sparse strata (to avoid having no unemployed Chinese-origin females in Wrexham, for example), which rather defeats the purpose. Despite its lack of popularity, judging from its rarity in the literature, random sampling is actually a lot easier than stratified sampling. And if the random sample is large anyway, as it should be, stratified sampling is not needed.

## Clustered samples

Perhaps one reason why simple random sampling is used so infrequently in social science research is that it produces a scattered sample (such as a few cases in every town). Where travel or lengthy negotiations for permitting access are involved in the fieldwork, **cluster random sampling** is therefore sometimes preferred. The cases we are interested in often occur in natural clusters anyway, such as institutions. So we can redefine our population of interest to be the clusters (institutions) themselves and then select our individual examples from within the clusters, whether using randomisation or any other approach. The institutions become the cases, rather than the individuals within them. For example, if we want to interview 100 nurses from hospitals face-to-face, we may not want a random sample of nurses from all hospitals in the country as this would involve too much travel. So we could select 10 hospitals at random instead and then interview 10 nurses in each. This would be a clustered sample, with the hospitals as 'clusters'.

This approach has several practical advantages. It is generally easier to obtain a list of clusters (employers, voluntary organisations, hospitals and so on) than it is to get a complete list of the people within all of them. If we use many individuals from each cluster in our selected sample, we can obtain results from many individuals with little time and travel, since they will be concentrated in fewer places. Even if we are not travelling we have to arrange memoranda of understanding to conduct the research, privacy notices, ethical clearance and so on, for fewer institutions.

Sometimes **cluster sampling** is the only way. There are occasions when it is preferable to treat groups of individuals as clusters. An obvious example would be the teaching classes in schools. It is theoretically possible to allocate individual pupils randomly to teaching interventions. But schools are much more likely to agree to take part if their existing teaching structure is respected, and whole classes are allocated to interventions as clusters.

One drawback of clustered sampling is the potential bias introduced to the sample if the cases in the cluster are too similar to each other. People in the same house may tend to be more similar to each other than to those in other houses, and the same thing applies to a lesser extent to the hamlets where the houses are (people in each postcode area may tend to be similar), and to the regions and nations where they live. This suggests that we should try to sample more clusters, and use appropriately fewer cases in each cluster. Another problem that is worse with clusters than with individual random sampling is that if one individual refuses to take part after being selected (the parent of a child in one class refuses permission to be in an experiment perhaps), then often the entire cluster has to be dropped. This causes much greater bias than if one individual drops out (Puffer et al., 2005).

Where clusters are used, they should generally and automatically become the units for analysis. If entire classes in schools are randomised to subgroups, then the allocation is at the class level. This is the level at which the analysis should be, thereby keeping the unit of analysis the same as the unit of randomisation (Cochrane, 2012). To do anything else makes the analysis more complicated for no analytical gain.

If hospitals are the clusters, and nurses are then selected randomly within hospitals, then we would have nested probability sampling (a cluster randomised sample). It is still perfectly proper to treat such a complex sample as a simple random sample of cases at the cluster level, and envisage the rest as repeated measurements. In fact, this approach has three key advantages. It is simpler, makes the subsequent analysis easier to conduct, and makes the analysis more robust, than treating it as a clustered random sample. For more on this, see Chapter 18.

## Exercise 10.3

Researchers want a good sample of hospital patients for face-to-face data collection, but they want to limit time and travel as far as possible. So, they decide on a multi-stage sample. First they select a random sample of the hospitals in the region of interest, and then they use all of the patients at each hospital.

a   What is the population and what are the cases here?
b   Is this a cluster randomised sample?

## Non-probability samples

If none of the above approaches is possible, then the research must be based on a **non-probability sample**. This is the most common, and overused, form of sampling. A convenience or opportunity sample is composed of those cases chosen only because they are easily available. A researcher standing in a railway station, or shopping centre, or outside a student union, and stopping people in an ad hoc manner would create a convenience sample, clearly not a random or representative one. This is so even if limits are used, to create **a quota sample**. It is not even clear that 'representative' means anything here, as there would be no easily defined population. Large numbers of people rarely travel by rail, shop in city centres or use a student bar. These people would tend to be excluded from such a sample. Those in paid employment may be less likely to be in shopping centres during the day, while older people may be less likely to go out at night. Convenience sampling introduces a very real danger of creating a biased sample. This should be a sampling strategy of last resort.

An example of a reasonable use of a non-probability sample is where a snowball technique is necessary. In some studies – of drug use, truancy, or under-age sex, for example – we are unable to produce a list of the population, even where the population of interest may be imagined as quite large. Indeed, one of our key research objectives may be to estimate the size of an unknown population. In such a project we might quite properly approach a convenience sample to get us started, and once we have gained their trust we could ask each individual to suggest other informants for successive stages. In this way, we hope that our sample will 'snowball'. Difficult-to-reach populations can make probability sampling impossible. We simply accept this, and do the best we can with what is available.

Non-probability samples can also be used when the population of interest is very small and we are approaching cases as expert informants. We may want to ask government cabinet members about the background to a new policy, or directors of large banks how they decide on investment priorities. In some studies the number of experts is so limited that we must use whoever is available to us and willing to take part, since there are not enough to select cases at random from a longer list.

### Exercise 10.4

A study of voting behaviour in a national election is based on researchers trying to complete a quick survey instrument with every 10th person who turns out to vote in 10 local polling stations. Around 80% of those asked did agree to complete the survey. Having finished the data collection, the researcher wishes to draw conclusions about national voting behaviour.

a   What is the sample and what is the population in this study?
b   This attempted generalisation from sample to national picture actually involves two rather different kinds of generalisation. What are they?
c   In the example described what are the most obvious limitations to each level of the generalisation from sample to national picture?

## HOW BIG SHOULD A SAMPLE BE?

We turn now to the required size of a sample. The size of a sample is important for its authority. All other things being equal (measurement accuracy, response rate and so on), the larger a sample is, the more convincing the ensuing study will be (see Chapter 8). So, make your own studies as large as feasible, and take account of the scale when reading the research of others. Too many studies are currently being conducted that have no chance at all of producing trustworthy results. This is a very common problem in stand-alone in-depth work, of the kind sometimes referred to as 'qualitative', but is also common elsewhere, for example in many psychology experiments. Sampling and deciding on a sample size are, like all aspects of research design, independent of the subsequent methods of data collection and analysis. If a sample of 200 cases is required to make believable comparisons between two groups, then 200 would be needed whether the cases were surveyed, measured, interviewed or observed.

Small samples lead to missing out on potentially valuable results (Stevens, 1992). If you are looking for a pattern among the data you have collected from a well-designed study, then your success or failure is chiefly determined by four things. These include the 'footprint' or effect size of the phenomenon you are studying. This effect size is often very small in social science – imagine the likely impact of a national budget policy on household happiness perhaps. And the strength of the pattern you are looking for is anyway not under your control, and not known precisely before the research starts (hence the need for research!).

How easy it is to spot a pattern also depends on its variability. If a change in happiness after a budget decision is small but is the same for all households then it is easier to detect than if it varies considerably between households. Imagine you were trying to find a difference between the average heights of two groups of people. If the people in each group are all of the same height then you only need a sample of one in each group to be perfectly accurate in your measurement, but the more variation there is in the heights of this population, the more people you need to measure to make sure the first few in each group are not extreme scores. The effect size you are looking for (difference between groups) would be small in comparison to the overall variability of your chief variable (height of individuals). This variation is also not under your control, and also not usually known precisely at the outset of the study.

The third factor is the quality of the measurements, and the absence of error, in your results (see Chapter 14 for more on this).

The fourth factor determining how easy it is to separate a pattern from the noise in your data is the size of your sample. This is the only element of success/failure (assuming a well-designed study) that is directly under your control, and which you can easily specify before you start. So, if using population data is not possible for any reason, make the sample as large as feasible. The sample must be large enough to accomplish what is intended by the analysis. Whatever you set out to achieve, data and cases will be lost (refusing to respond and so on), and so your achieved sample will probably be smaller than you intended. This is another reason to make the sample as large as possible from the outset. However, there is no clear technical answer to exactly how large that should be.

## Deciding on your sample size

Two factors you could consider in selecting a sample size are the number of variables you expect to use in the dataset from your sample, and the number of comparisons you will make between subgroups of your sample.

For many of the multivariate analyses described later in this book it is assumed that the number of cases in the analysis is far larger than the number of variables. Some methods resources might suggest at least 10 times as many cases as variables. Others might insist on 20. I might prefer hundreds. The point is the same though – do not set your sample at anywhere near the same order of magnitude as the number of issues you want to consider or take into account. And remember that the number of variables is liable to increase throughout your study (e.g. Chapter 17 describes the practice of creating several 'dummy' variables from each categorical one).

Perhaps even more relevant is the number and range of comparisons you intend to make during your analysis. The quality of your study sample should be judged not on the basis of its actual size, but on the number of cases in the smallest cell for any comparison (see Chapter 8). Having a sample size of 100,000 cases is of no use if your intention is to compare the incomes of residents in the north and south of a county, and 99,999 cases are from the south. A cell size of 1 (from the north) makes this a trivial study even though the overall number of cases is large. So, a simple plan would be to work out how many cells there would be in the most complex comparison that you want to make. This is another reason for having clear research questions and an explicit research design before you start, so that you know beforehand what the most complex comparison will be (Chapter 8).

If your most complex comparison is between the average incomes of residents living in the north and south then you have two cells, based on one comparison variable with two values – north or south. If you want to know whether the parents of ethnically white boys are more likely to agree with their child's aspirations in life than parents of girls, or of boys from other ethnicities, then you have two key comparison variables. These are the ethnicity and sex of the child. This means that your analysis will have at least four cells (white boy, other ethnicity boy, white girl, other ethnicity girl). If you want to compare the agreement between parents and their children's aspirations for each of seven ethnic groups (white UK, white other, Black African origin, Black Caribbean, East Asian, South Asian, and other, for example), then you have 14 cells – one cell for each combination of ethnic group and sex of child. It is remarkable how quickly the number of cells expands in real-life research. Try drawing it up as an empty table. Put your imagined sample size as the total in the table, and work out the 'expected' size for each cell (as explained in Chapter 3).

If you cannot work out how many cells you will use, this may suggest that you have not planned your research properly.

Once you have an idea of how many cells you need to answer your research questions, consider how small the number of cases could be for the smallest cell. How low could this go and you would still trust the research? In comparing the average incomes of north and south, would you be happy with 10 cases in each cell, 100, or more? Multiply this minimum cell size by the number of cells, and this gives you the bare minimum for your required sample size. So, if you decided on a minimum of 100 cases per cell and had 14 cells in your comparison (as above) then you would need a minimum of 1,400 cases in your sample. In reality you would need many more because although you might assume that the split of boys and girls will be near equal, you cannot assume the same for ethnic origin. In the UK there would be more children of reportedly white origin, and fewer of reportedly East Asian origin. To obtain a sample with at least 100 East Asian origin children of each sex you would need far more than 1,400 cases in total.

In summary, whether you will successfully find a pattern or trend (or none) in your dataset depends on the strength of that pattern in reality, the variability of the cases, the quality of the measurements, the size of the sample, and the number of comparisons you make with it. The first two are beyond your control. An important part of your research design should be to ensure that the expected minimum number of sample cases in each cell of any comparison in your study is sufficient. 'Sufficient' here means something like 'convincing to a sceptical reader'.

## Reasons for using a smaller sample

There are, of course, some situations in which there are ethical reasons for not having a particularly large sample. The main example would be when you are conducting an evaluation with an intervention. If there is a potential danger from the intervention, it makes sense to minimise the possible harm to cases in the study by limiting their number for the first trial.

In health science, for example, a study may be the first test of a new medicine on human (or animal) subjects. Here, the test is for effectiveness but also for possible side effects. In this situation, it would be best to conduct the test using the smallest number of cases that could give a convincing demonstration of the use and safety of the medicine. There will be many analogous situations in social science. New rules for banking, new qualifications for school-leavers, and new procedures for benefits payments should all be tested before being applied more widely. The test should be for effectiveness, relative cost, and for any unintended consequences. As with testing medicines, ethical considerations might suggest that the test should be rigorous, but initially conducted using the smallest possible number of cases needed to be convincing.

## How not to pick the size of a sample

Having limited resources to use a sample of the scale that the research demands is not a good reason for limiting the size of the sample. If you do not have the resources to do the research properly (so that the results can be trusted) then do not do it at all. Setting out to do research badly is unethical, and wastes everyone's time (yours, the participants' and the readers').

The appropriate scale for a sample is also not closely related to its size proportionate to the population. It does not have to be a certain proportion of the population. If a sample of 100 cases were considered appropriate for a particular study, it would be appropriate whether the population was 1,000 or 1,000,000.

When using a random sample, there is still a tradition among some researchers of conducting a 'power' calculation to determine a minimum suitable sample size to detect an effect of a particular size (Chapter 7). A simple approximation of this calculation is to have a minimum of 16 cases, divided by the desired effect size squared, in each subgroup of the sample, where subgroups are compared to estimate the effect size. For example, if observations of males and females in the sample were to be compared, with an estimated effect size of 0.4, then the number of cases per group would be 16/0.4$^2$, or 100. The total sample would need to be 200 cases, because there are two groups in the comparison. This would, in the logic of power calculations, give an 80% chance of detecting an effect size of 0.4, at a significance level of 5%.

There are several problems with this kind of calculation. It is circular, very sensitive to minor changes in assumptions, and depends on having a good idea of what the research will find before it is conducted. But most importantly, the whole approach is based on the archaic approach of significance testing. Significance testing does not work as intended (Chapter 6), and so power calculations predicated on significance testing do not make sense for the same reasons.

## CONCLUSION

This chapter looked at how to decide on the cases for your study, presenting different approaches in an approximate order of preference. Population data is usually best. If this is not possible then a large, true random sample is the next best. Variants such as systematic and cluster random sampling are worse, and a non-probability sample is usually worse still in terms of the potential for bias. The choice you make should be justified and well argued in your research report.

The first and main purpose for having a high-quality set of cases to address your research question is so that the results you get are (internally) valid. Whether they are also true of a wider set of cases not in your research is a secondary issue, which is relevant only if your results are valid. The easiest way to address both the validity and generality issues is to use population data. Whether you use a population or a sample, both will usually be incomplete and therefore have some potential for bias. This means it is safest to treat the cases in *any* study as a kind of incomplete population in the first instance, check for internal validity, and worry about generality later, and only if it is relevant because the initial results are deemed valid.

### ━━━━━━ Notes on selected exercises ━━━━━━

### Exercise 10.2

The population consists of all nurses working in that hospital at the time (i.e. everyone who had a chance to be in the study). If there were 30 nurses in the sample, instead of 20, the population would remain the same - all of the cases that could be in the study (unaffected by how many actually are in the study). If the study were based on a survey rather than interviews, the population would still remain the same. The elements of research design, like the cases used, are independent of the methods of data collection.

If the study involves patients as well as nurses this does change the population, because it changes who had a chance to be part of the study. Either the population is now all nurses and all patients present in the hospital, or there are two populations – of nurses and patients. This largely depends on whether the researcher wants to treat them as one group or two.

## Exercise 10.3

a   The cases are the hospitals, since it is these and only these that are being selected. The population is the set of all hospitals in the region (i.e. all of the cases that could have been selected).
b   According to the interpretation in this book, this is not a cluster randomised sample, if that means there is randomisation at more than one level. It is a relatively simple random sample of hospitals. The only sampling variation that exists is at the hospital level. The research is then a census of each hospital. To be a cluster randomised sample, the researchers would have had to pick patients or wards at random within each hospital chosen at random. The cluster sample actually used here is preferable, if resource allows, to the cluster randomised sample since it eliminates sampling variation at the within-hospital level. However, these terms are used in different ways in different fields, so take care to look beyond the name.

## Exercise 10.4

a   The population in this study is all voters who turned up to the 10 local polling stations, and the achieved sample is the 80% of every 10th voter who actually took part in the survey.
b   The researcher is trying to generalise from the sample to the population, as standard, and is then trying to generalise from the population of 10 polling stations to all polling stations in the country. The individuals are a sample of the 10 polling stations which are, in turn, a kind of sample of a larger national population of voters.
c   Although taking every 10th attendee is not ideal, and may be impractical at busy times, it should give a reasonably true picture of voters throughout the day. The bigger problem is the 20% of voters who refused to complete the survey. We have no reason to believe that this 20% is the same in all other respects as the 80% who took part. As ever, the non-respondents could be busier, less confident, or even slightly embarrassed by their voting behaviour. This is a serious limitation to the generalisability of the sample, and should be mentioned in any report of the research. It will reduce how convincing readers will find the results. The attempted generalisation from the 10 polling stations to the national picture is even more problematic. The example does not state how the polling stations were selected but they are clearly not a fair representation of a whole country. The number of cases is small, and they are local to one area. These limitations will inevitably make the local picture less convincing than if the response rate was 100%, and make the national picture less convincing than if a somewhat larger number of polling stations had been chosen at random across the country. However, this is quite normal in real-life research. None of these limitations is fatal, as long as the researcher does not try to hide them, or to pretend that they can be overcome with statistical jiggery-pokery.

## Further exercises to try

Using any dataset with at least 100 cases, create a random sample of 20 cases. Do this by any fair means. For example, using the random number generator in Excel, create a new column (variable), and generate a uniform random number for each case. Then pick the 20 cases with the smallest associated random numbers.

Run a simple descriptive analysis for any other variable in the dataset (as in Chapter 3), but just using the selected 20 cases. Note the summary result (mean, mode or whatever), and then randomise 20 cases again. Are any of the second batch of cases also in the first batch? Rerun your analysis and note how similar or different the summary result is between the two samples.

## Suggestions for further reading

An introductory book on sampling:
Henry, G. (1990) *Practical Sampling*. London: Sage.

A chapter on the role of populations and samples:
Chapter 6 in Gorard, S. (2013b) *Research Design: Robust Approaches for the Social Sciences*. London: Sage.

A book on more advanced sampling techniques:
Thompson, S. and Seber, G. (1996) *Adaptive Sampling*. New York: John Wiley & Sons.

# 11
# WHAT IS RANDOMNESS?

## SUMMARY

This brief and more conceptual chapter describes the elusive idea of a random event, and what a set of random numbers is. The chapter shows that it is just about impossible to tell whether any sample is truly random, and illustrates the difficulties of creating a truly random sample in practice. These ideas have implications for what kinds of analysis are, and are not, feasible in social science. New researchers may want to skip this chapter, and move on to looking at handling missing data in Chapter 12.

## WHAT IS RANDOMNESS?

In statistics and the philosophy of science, 'randomness' means something rather more than its everyday meaning of merely haphazard, or without apparent pattern. It is stronger, in the sense that something either is or is not random (Gorard, 2013b). There are no degrees of randomness, because a random event is one that is completely unpredictable in form, outcome or timing. Randomness is the quality of such an event – its unpredictability, lack of intention and lack of long-term patterning.

Applied to a set of numbers, randomness means that knowing some of the numbers in the set will not help you to identify any of the other numbers in the set. By analogy, if a standard six-sided die roll has a random outcome, then knowing the previous 10, 100 or 1,000 results from that die will not assist you in predicting the next one. The chances of guessing the result of the next roll correctly remain at 1 in 6, however many rolls have been seen previously. Randomness is the characteristic of chance, as illustrated by a fair die.

Of course, there can be short-term patterns in random events, just like shapes in the clouds. A six-sided die can roll a six 10 times in a row and still be random. In fact, if this never happened its randomness might be doubted. Random events can have a clear distribution of probability over a large number of events, and this is a kind of pattern. For a die roll the distribution is called 'uniform' because each of the six outcomes has the same likelihood (distributions are discussed in Chapter 3). So, over a very large number of rolls, the outcomes will tend to be equal in occurrence. There will be something like 1 in 6 ones, 1 in 6 twos, and so on. But this does not help to predict the next random event (to believe otherwise, and that the die 'owes' a 6, for example, is known as the 'gambler's fallacy').

There have been various attempts to define randomness more formally than this over time, in terms of unpredictability, and the long-term probability of proportional selection from population characteristics (von Mises and Doob, 1941). The most promising approach is perhaps in terms of the non-computability of random events. For example, Kolmogorov and Uspenskii (1987) define a set of units as occurring randomly if the set cannot be described more efficiently than by repeating the entire set. Randomness, and whether such a thing even exists, or indeed whether everything that happens is random, are discussed in much more detail in Gorard (2013b). How to generate random numbers is illustrated in Chapter 3.

In a sense this is all that needs to be said here. But randomness and randomisation are such key ideas for sampling, handling missing data, allocation to treatment groups in experiments, and statistical analysis, that it is worth considering them a little further.

**Image 11.1** The die is cast

## PROBLEMS WITH RANDOMNESS

In real life, truly random events will be rare. It is hard to imagine a die has been so perfectly manufactured and weighted that each side will have precisely the same probability of occurrence. It is also possible to imagine that a person rolling even a fair die might be influenced slightly by the faces showing when the decision is made to release the die. Even random number tables generated by a computer must be based on the algorithm (the program) that creates them, and this creates the danger that they are not truly random. In real life, the results of events such as rolling a die or 'computing' a random number are therefore termed 'pseudo-random'. They appear random and they are as good as random in practice (i.e. unpredictable). Randomness is more of an ideal to aim for than something that can be made to happen in practice.

People are not good at judging and understanding randomness, and cannot identify it accurately based on a finite sequence of data units (Volcan, 2002). Of course, non-random cases can be more or less predictable, and easier or harder to compute, but this is not what makes them random or not. For example, the sequence 1, 2, 3, 4, 5, 6, 7, ... is slightly easier to imagine as a pattern, and to predict the next item in, than the equally patterned sequence 1, 4, 9, 16, 25, 36, 49, .... The second sequence is just the square of the first. But this does not affect whether either sequence is random – something which it is impossible to tell in practice, or by inspection. The first sequence is just as likely to occur by chance as the second one is, and has the same probability as the sequence 2, 5, 6, 4, 3, 1, 7.

## Random sampling

A 'random sample' is a subset of cases from the known complete population (see Chapter 10), selected by chance in such a way as to be completely unpredictable. The chance element can be produced by observing radioactive decay, using specialist software or Excel, via a random number table, or a mechanical process like a card shuffling machine. Strictly speaking, these all produce only pseudo-random numbers, because they are generated by a process of some kind, and one cannot *cause* a random event by definition. But they are all that is possible in reality.

A true random sample should also permit the occurrence of the same case more than once (like rolling a die or drawing a card from a pack of 52, replacing the card, and drawing another one at random). The chance of such repetition depends upon what proportion of the population is in the sample. If the population is large in relation to the sample then this issue of permitting repetition within a sample may not matter much in practice, but it is important to remember that sampling *without* the possibility of repetition (of the card or whatever) is not really random sampling. So, a 'random' sample is really pseudo-random, and if it was drawn without the possibility of repetition then it is even further from the ideal of randomness. Compromises like this, however small, mount up and mean that real-life samples are rarely if ever random.

Most importantly, a random sample must be complete in two senses. It must include every case that was selected from the population by chance. There can be no non-response, refusal or dropout in a random sample. And every case selected must have a known measurement or value of the characteristic of interest (height, income or whatever). There must be no missing values. If any cases or values are missing then the sample becomes non-random by definition.

## CLUSTER RANDOMISATION

How does randomness relate to cluster randomisation, as described in Chapter 10? It should be clear, whichever definition of randomness is used, that cluster randomisation is not the same as being truly random.

A simple cluster sample is one where the unit of sampling is a cluster, such as a hospital, even if data is collected from more than one individual (or ward) within each hospital. This sample is a pseudo-random one, and can then be analysed as a sample of hospitals.

Imagine we wanted to measure very accurately the heights of the people at an academic conference. Our population could be the people at the conference, and our sample might be 100 of the conference delegates selected at random. We might check these delegates' heights with several instruments and at several different times of day, and then average the results for each person. In this example, our measurements would be repeated for each individual in order to maximise accuracy. And our outcomes would be the averaged heights for each person. So is our sample really the people, or is it the separate measurements of heights? Imagine each of 100 people had their height measured four times. Is our sample size 100 or 400? Traditionally, this would be treated as a sample of 100 cases with four repeated measures for each. This makes sense. The sample is defined by the level at which the randomness occurs, in the selection of the people. There is no randomness or probabilistic uncertainty in terms of how many or what types of measurements were taken for each person. The repetition of measurements within the 100 cases is to increase the accuracy of the estimate for each case. It does not increase the sample size.

A **cluster randomised sample**, on the other hand, is one where there is more than this one stage of randomisation. For example, a sample of hospitals is selected at random first, and then some patients are selected at random in each hospital. Imagine a study of patient satisfaction in hospitals. The researcher selects a random sample of 10 hospitals, and then (and only then) a random sample of 100 patients within each of those 10 hospitals. What is the sample size here? Is it 10 hospitals or 1,000 patients? Both answers are possible. If the concern was primarily to compare variation in patient satisfaction between hospitals, then $N = 10$, and the repeated measures of individual patient satisfaction within each hospital are there to create a better estimate for each hospital (just as with the height example above).

On the other hand, if the purpose was to estimate patient satisfaction in hospitals in general, and the first step of selecting hospitals was for convenience (otherwise the sample could inconveniently consist of a few patients in all hospitals in the country), then $N = 1,000$. But this would no longer be a random sample. This is somewhat confusing. The answer really depends upon your research question and the purpose of your analysis. You could never go wrong by treating it as a sample of 10 (see Chapter 17 for more on the analytical implications, or their absence, when using a cluster random sample).

## CONCLUSION

Given that randomness is an ideal not often seen in practice in social science samples, it is most important that the much more common non-random (and incomplete 'random') samples are analysed properly. And that appropriate techniques are available and taught for handling these real-life samples that are not random (but which are largely ignored in most statistical texts). Real-life samples cannot and should not be used with any technique predicated on random samples. What this means for missing data is discussed in Chapter 12, and what it means for conventional inferential statistics is explained in Chapter 6.

### Suggestion for further reading

A wider discussion on randomness, and whether it is possible:
Chapter 5 in Gorard, S. (2013b) *Research Design*. London: Sage.

# 12

# HANDLING MISSING DATA: THE IMPORTANCE OF WHAT WE DO NOT KNOW

### SUMMARY

In any real-life dataset, non-response is not randomly distributed. There are proven systematic differences between people who tend to take part in research and those who refuse – in terms of leisure, attitudes, education, income, social class, age and so on. Whether they are people or organisations, all of the cases in any study are effectively volunteers, and these willing participants could be very different from those who are not. Even among those who do respond, there will be questions unanswered, or unintelligible responses. And these forms of missing data have also been shown not to occur by chance, but lead to potential bias in the results that are achieved in any study. This chapter looks at the important issues of how to detect missing data, how to report it, and how to deal with it in analyses. The next chapter discusses some of the more technical issues underlying the guidance in this chapter, and more can be read about this in Gorard (2020).

### ILLUSTRATING THE PROBLEMS OF MISSING DATA

Imagine that you are trying to estimate the average annual income for all residents in one town, and that there are no up-to-date official figures on income, so you have to collect the data for yourself. You do not have the resources to collect the relevant data from all residents (perhaps there are around 100,000 people in the town). So you decide to use a representative sample of 1,000 people (1% of the total).

In order to select a truly representative 1% sample you would need to start with a list of all residents in the town. The ideal would be to select 1,000 cases randomly from this list, perhaps by giving each resident a meaningless identity number, and using random number generating software to select 1,000 of these identity numbers. Excel will generate random numbers, as will SPSS and other analytical software, and there are many bespoke apps online that will do the same (see Chapters 3 and 10). In theory, this would provide an unbiased subset of all residents on the list.

In practice, the subset of residents generated by this approach depends on the quality and completeness of the full list of residents. However recently it had been compiled, the list is likely to be out of date. People will have moved in or out of the town since the list was completed. Some will have been born, and sadly some will have died. This would make your otherwise random sample biased (as a representation of the town, whether at the time that you completed your study or when the list was compiled).

## Exercise 12.1

Why does data missing because your list is out of date probably mean that your random sample will be biased?

---

On average, those people who have moved in or out of the town are likely to be more mobile for a reason that could be related to their income. They may be more likely to be in national professions that are more highly paid, or more motivated and ambitious job-seekers, or students moving to university without substantial income. Maybe they moved to live with a new partner, perhaps because they are younger than the average for the town. Your sample may underrepresent those born most recently, with no income, and it may also underrepresent the elderly and those with serious ill-health who may have lower than average incomes. Your list of residents may well omit many of those in the town who are homeless, who are also likely to have lower than average incomes. Others may be away on holiday, or ill in hospital, or maybe even in an institution such as a prison. Those on holiday may have somewhat higher incomes, and those in institutions somewhat lower incomes.

All of these examples will lead to missing cases in your study, and whenever the reason for any cases being missing is related to the topic of the study (here it is income), then this will inevitably lead to bias in your findings about that topic. The examples so far may have tended to remove cases with low or high incomes, so making the remaining cases appear to have less variable incomes than in reality. Note that these problems cannot be overcome by any other method of selection or sampling (we will consider later in the chapter what might be done to help address such problems). If you do not have a full list of residents, then your sample is bound to be biased. If you do not select randomly from the list, but instead pick cases in terms of residents' known characteristics, or for convenience, or arbitrarily, then each of these approaches is bound to produce an even worse sample (see Chapter 10).

**Image 12.1** An empty chair

Unfortunately, though, the problems have only just begun. The stage of picking a sample is under your control, and can be as high quality as you are prepared to make it (see Chapter 10). But merely picking cases does not mean that all will respond positively, if you then try to ask them about their annual income, for example. The biggest problem in a study like this will be non-response. Some, perhaps most, of the cases selected and seemingly available to respond to your questions, will be unwilling or unable to answer.

Partly this depends on how you try to collect the data. If you use a written survey, then it is unlikely to get a response from residents with poor levels of literacy. If instead you visit addresses to talk to people then you are likely to miss those who are at work when you call round, or who do not answer the door because they fear you are a debt collector (a real example that has happened to me). If you use an online survey then you are excluding residents without easy access to a computer, smartphone, or the internet, and so on. All of these factors will be related to the incomes of the people missed out by each approach.

The single largest group of missing cases will probably be those who simply do not want to answer the question about income. Even in the official cohort study in England known as Next Steps, for example, only 42% of respondents provided a valid estimate of their income in the first round of data collection, and there is clear evidence that this subgroup of responders yielded a biased estimate of income (Siddiqui et al., 2019).

Many people will not know their income precisely. Even people who do know what their income is may not want to state it. They may be embarrassed, feeling that is very high or very low, they may not want tax and pensions officials to know about their extra investments, or they may be working unofficially and being paid in cash to avoid tax altogether. Even if you assure residents that their responses will be treated as confidential and that they will never be identified, it is probable that many will still decline to respond. In sum, these factors are likely to lead you to a misestimate of the true incomes of the town residents. This is further clear bias.

All of these issues, or something like them, will occur in any real-life research, collecting any kind of data. There will be missing cases from samples and even from populations, there will be cases who do not respond, and there will be measurement error in the results even from cases who do respond (such as when respondents provide the wrong income for whatever reason). This is partly why approaches to analysis such as significance testing are so rarely, if ever, justified (Chapter 6). Significance tests require full randomisation, no dropout and no measurement error – a perfectly acceptable set of assumptions in maths or theory, but completely unrealistic for social scientists.

### Exercise 12.2

What other reasons can you imagine for data being missing in any dataset? Be imaginative and discuss.

## WORKING THE EXAMPLE

To see how this might all work out, assume that the average annual income for the town of 100,000 people was really £30,000 (although we would not know that, else we would not be doing the research). Assume also that if we had a full list of residents, an unbiased random sample of 1,000 people, and all agreed to respond and did so honestly, then our sample estimate would also be £30,000 (or very similar). This is our ideal target – to get a sample result that is as good as a result from the whole population.

Perhaps 5 of our sample of 1,000 chosen cases were no longer resident and so were out of contact when we did the study, a further 95 cases were not contactable by the methods we used in the survey (no address or wifi perhaps), and a further 200 refused to answer the question about their income. We are now missing values from 300 cases in our ideal scenario. Given that these 300 cases are unlikely to have an average income of £30,000 as a group, the results from our actual achieved sample of 700 are now biased. We could end up with fundamentally incorrect results for our research.

The achieved sample may underestimate the average income for the town. If the average income of the 300 missing cases were actually £50,000 (unknown to us of course), then our sample of 700 would suggest that the average income for the town was nearer £21,000 (because 300 times 50,000 plus 700 times 21,429, all divided by 1,000, is 30,000). This figure of £21,000 would be a gross underestimate, and could distort any decision predicated on this supposed 'result'. The same kind of bias would occur if income were overestimated, or even if only the variability of income was obscured by missing many cases with high and low incomes. The latter would bias any ensuing analyses, such as effect size calculations (Chapter 5).

Bias is dangerous, and not just to the validity of research findings. Bias has real-life consequences. For example, Her Majesty's Revenue and Customs in the UK may release a national income figure of £30,000, based on more accurate but confidential tax returns, unavailable to you. This would suggest that our town is relatively low earning at £21,000 (as above), based on inaccurate figures from our new research. This may persuade policy-makers to invest more heavily in this town at the expense of investment elsewhere, and so reduce investment in other areas that genuinely *do* have low average income. Even large-scale, careful research can lead to apparent implications that are in fact completely unfounded, and so create damaging consequences for real lives. The trustworthiness of research findings (see Chapter 8) is an issue for research itself, but often also practically, and for social justice and ethics.

## Exercise 12.3

Using the same example, with a true figure of £30,000 for the full intended sample, what would your achieved sample of 700 cases have as its average income if the missing 300 cases had an average income of zero? If you get the answer without using a calculator, award yourself double marks!

The calculation above would work like this. The mean income of the achieved sample ('??' in Table 12.1) must be such that the overall mean income of 30,000 is correct. So 700 times ?? divided by 1,000 must equal 30,000, meaning that ?? is £42,857 (30,000/0.7). In this example, based only on the achieved 700 results, we would grossly overestimate the average income for the town.

Table 12.1  Estimating the income of an achieved sample

|  | Achieved sample | Missing cases | Overall known figure |
|---|---|---|---|
| N | 700 | 300 | 1,000 |
| Mean income | ?? | 0 | 30,000 |

## COMPARING TWO OR MORE VARIABLES WITH MISSING DATA

However, most social science research is more ambitious than simply estimating the average income for one town. As soon as the research becomes even a little bit more complex, then the bias created by missing data gets worse. For example, a simple research question might involve comparing the average income for two or more groups (by age, gender, ethnicity, and so on). This introduces at least one other variable, such as age, as well as income. And this again raises the possibility that people cannot or will not provide information on their age accurately and fully. If we collect income data but no age from one resident, and age but no income data from another resident, then we cannot use either case in our analysis. Adding another variable generally adds missing (and inaccurate) values for more cases in the sample, and so reduces the size of the effective sample, and also tends to increase the bias further. Most studies use many variables, and each one is likely to increase the problems of bias in the achieved data.

Another simple research question might be about how much the incomes of the residents in the town have changed over time, perhaps before and after some other event such as a recession. Here the data on income needs to be collected from the same people on at least two occasions. We will have to ignore any cases where we have a response about income on the first occasion but not on the second (we obviously cannot use only one value to compare change over time). And we also have to ignore any cases where we only have a response the second time. People may have moved to or from the town, been away on the first occasion and not on the second, or vice versa. People may have changed their mind about whether they should respond. This attrition or dropout of cases, like all of the examples of missing data so far, provides an opportunity for bias if the reasons why attrition exists are related to the variables being collected in the study.

---

### Exercise 12.4

How might missing data occur or become an added issue if you wanted to compare the average income of the residents in two towns - the one involved in the example so far and another one?

---

## SUMMARY SO FAR

In almost all large-scale or long-term social science research, there will be data missing from the dataset to be analysed (Berchtold, 2019). This could be because of refusal, or non-response to an invitation to participate, leading to entire cases being missing from the dataset at the outset. It could be due to dropout or attrition, where cases that had agreed to participate changed their mind, or were prevented by circumstances from providing data. Attrition may be due to mortality, illness, or the busyness or mobility of the cases involved. Each situation leads to entire cases unavailable for substantive analysis, by the end of the study. All non-response creates a potential for bias in the results based on analysing the remaining cases (Peress, 2010). This would lead to invalid results, and would affect the generality of any findings to cases that had not been involved in the study (see Chapter 8).

There will also be values missing from variables for the cases that *are* in the dataset. This could arise from partial non-response, refusal to answer an intrusive or complex question, indecipherable responses, loss of data, incomplete coverage, errors such as inadvertent skipping of question items, and a range of related reasons. However, the issues raised so far are not limited to large-scale work or to research involving numbers. They are probably most damaging in the common kinds of small social science study that take no account of missing data at all, and therefore whose validity is always in serious doubt. Again, it is very unusual to find a social science dataset of any kind in which every case has complete information. Full participation in any social science study is so rare as to be unheard of (Cuddeback et al., 2004; Lindner et al., 2001), and this would affect the trustworthiness of the achieved results (Gorard et al., 2017).

In combination, missing cases and missing values mean several things for any analysis we then carry out. It is safest to assume that all datasets are incomplete. This also means that cases are never truly randomly selected, or randomly allocated to groups, because a random sample which is incomplete is no longer random (Hansen and Hurwitz, 1946; Sheikh and Mattingly, 1981). Any form of analysis predicated on randomisation cannot be used where any data is missing. Missing data also always creates the potential for bias, as illustrated above, and this can have an important influence on research findings.

### Exercise 12.5

Look at an empirical research study (paper or report) that you want to read anyway. Does it specify where and how any data is missing? Is this clearly described? If not, why do you think the author has not been clear?

### Exercise 12.6

A new study has conducted 500 home-based structured interviews with householders. There were 300 middle-class families (according to their responses) and 200 working-class. Because of refusals, 1,200 homes were approached to gain these 500 responses.

a   What is the response rate in this study?
b   How different from this 60:40 distribution of the classes would the non-responding families have to be for there to be actually more working-class families?
c   Assume that the study does not attempt to generalise, but is only concerned with the achieved sample of 500. Imagine also that the process of deciding who is middle or working class leads to around 10% misclassifications in each direction. The study reports that middle-class respondents are more likely to own a car (200 of them) than working-class respondents (of whom only 100 had a car). How safe is this claim, considering the cases in the study?

## PREVENT MISSING DATA AS FAR AS POSSIBLE

However, all is not lost. There are practical steps we can take to minimise missing data, and deal with it when it occurs.

The most effective and safest method for handling missing data is to try and prevent its occurring in the first place. Once you have selected your ideal sample, one of your main priorities should be to achieve that sample, and minimise any non-response. This is what a lot of the craft of conducting research is about. Design your study to make it possible to get as near to 100% response rate as you can, by making access and participation easy, brief and rewarding. Do not ask for data that is already available elsewhere. Make sure the research is important, and interesting to participants, and not a waste of their time. Make your research instruments easy to read, and complete, avoiding jargon and those long words and sentences that social scientists are prone to. Make the questions clear, non-threatening and unobtrusive. Make the first question fascinating. Ensure that any instruments are accessible to their full intended audience, including those with limited literacy, reduced visual acuity, and so on. Follow up missing responses, chase up non-responders, and treat every response with care and appreciation. Make clear that all data will be treated with respect, respondents will be anonymised, data will be destroyed after use, and offer any other reassurances. Data is valuable.

If you are conducting research with a design such as a randomised control trial (RCT) or longitudinal cohort study (Chapter 9) that involves participation over time then the issue of missing data through participant attrition can be even more serious. There is a range of steps to take. In a **placebo** experimental design, both the group randomised to receive the intervention and the control group who do not receive the true intervention will actually appear to have the intervention. One is given the real intervention and the other group gets a pretend intervention (or placebo). More realistically in social science, a crossover is possible, whereby each group acts as the control group for an intervention used with the other group. Even better, simpler and more ethical is a wait-list design whereby all cases receive the intervention, and randomisation is used only to decide exactly when that happens (Gorard, 2013b). All of these formats are intended to reduce the possible demoralisation of participants in the study once they learn which group they have been allocated to. In each format all cases take part in an intervention of some kind.

Also, in more complex or lengthy studies, participants should be made aware from the outset of the difference between being in the study and being part of any intervention. This means that if they walk away from the intervention (or control) they realise that their input to the study itself will still be valuable by providing data. This can be achieved partly by asking each participant to agree a memorandum of understanding (MOU) which lists the rights and responsibilities of all parties. Further discussion of these issues appears in Gorard et al. (2017).

### Exercise 12.7

You know that getting responses from Travellers and homeless people, especially if you need their help on two occasions, will bring your response rate down. So unless learning more about the homeless is a specific purpose for your research, maybe you should plan not to ask about residents but about home-dwellers. By redefining your question you have evaded at least some loss of data. Or have you? Discuss. What problems might you have introduced instead?

## REPORT MISSING DATA ACCURATELY

A good next step is to acknowledge and record all forms of missing data, the reasons for its being missing (if known), and the stage of the research at which it occurred. This then needs to be reported clearly. Many research reports omit the response rate and how missing data and cases were handled in the analysis. Many studies present different numbers of cases in different sections of their results without explanation or audit trail. All of these are incorrect reporting. A research report should make the following information available, if known:

- How many cases were assessed for eligibility to take part in the study
- How many of those assessed for eligibility did not participate, and for what reasons (not meeting criteria, refused and so on)
- How many agreed to participate
- How many were allocated to each group for any comparison (if relevant)
- How many were lost or dropped out after agreeing to participate (and after allocation to a group, if relevant)
- How many were analysed in any step of the analysis
- Why any further cases were excluded from the analysis.

A useful format for many studies is a CONSORT flow diagram (Consort, n.d.). A simple flowchart for the income survey example used in this chapter might be as in Figure 12.1.

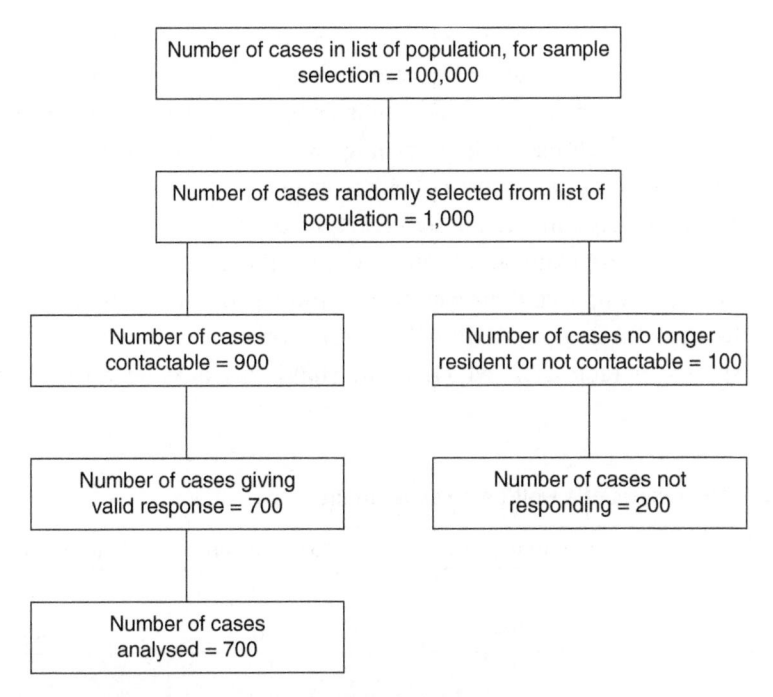

```
┌─────────────────────────────────────────┐
│ Number of cases in list of population,  │
│ for sample selection = 100,000          │
└─────────────────────────────────────────┘
                    │
┌─────────────────────────────────────────┐
│ Number of cases randomly selected from  │
│ list of population = 1,000              │
└─────────────────────────────────────────┘
         │                      │
┌──────────────────┐   ┌──────────────────────┐
│ Number of cases  │   │ Number of cases no   │
│ contactable = 900│   │ longer resident or   │
│                  │   │ not contactable = 100│
└──────────────────┘   └──────────────────────┘
         │                      │
┌──────────────────┐   ┌──────────────────────┐
│ Number of cases  │   │ Number of cases not  │
│ giving valid     │   │ responding = 200     │
│ response = 700   │   │                      │
└──────────────────┘   └──────────────────────┘
         │
┌──────────────────┐
│ Number of cases  │
│ analysed = 700   │
└──────────────────┘
```

**Figure 12.1** A flow diagram to record missing cases

Of course, in reality it is very unlikely that the numbers would be as neat as this. In this example, we could summarise the situation by saying that missing data accounted for 30% of the originally planned sample. This is a high and potentially fatal proportion for a representative sample, but it is actually less than you will tend to encounter in much of the research that you read – assuming that the author bothers to report missing data at all. The vast majority of social science research currently does not report these steps, making it impossible to judge how trustworthy the results are (Chapter 8). As a rule of thumb it is best simply to ignore (or at least give much less weight to) research that does not report missing data properly.

## ANALYSIS OF MISSING DATA

The next step is to calculate and report whatever can be found out about the data that is missing. It may seem odd to suggest analysing the data that we do not have, but in fact quite a lot can be done in this respect. It is possible to run descriptive analyses of anything that is known about the cases that are missing data (see an important real example of this in Chapter 4). Sadly, there is much less we can do about the cases that are entirely missing, and about which we know nothing (but see below).

Imagine that we have a dataset including variables for both age and income. We can reasonably calculate whether the cases missing only a value for age have a higher or lower than average income. The answer tells us something about who is missing age data. We can also calculate whether the cases missing income but not age are older or younger than average. Again, the answer tells us something about the missing cases for income. This approach may help us to judge whether the likely bias in our sample will lead to an over- or underestimate of either value.

For example, in an experimental design (Chapter 9) we might have agreed an MOU with 400 people to participate, and then randomly allocated 200 to receive an intervention intended to improve their well-being, and the other 200 to act as a control group. At the outset we surveyed all 400 in terms of their well-being, and obtained the results shown in Table 12.2, which shows standardised scores for well-being out of 100.

**Table 12.2** The results of a well-being experiment

|  | Intervention group | Control group | Overall known figure |
| --- | --- | --- | --- |
| Well-being score | 73 | 71 | 72 |
| N | 200 | 200 | 400 |

If the overall standard deviation of the well-being scores was 24, then a difference of 2 points (73 – 71) is equivalent to an effect size of +0.08 (or 2/24). At the outset the intervention group was already ahead of the control by a small amount, and this would have to be taken into account when considering the final outcomes. Imagine next that 20 cases dropped out of the study from each group, and did not agree to provide a well-being score after the experimental intervention. This is 10% overall attrition, and equally balanced between the two groups. Some commentators, and even purported authorities such as the What Works Clearing Houses, would call this 'low attrition', of the kind to be expected in any study.

However, because we have the pre-intervention scores for all 400 cases we can examine the initial scores for the 20 dropouts in each group. If these scores are as in Table 12.3, then there actually is a problem for the study. It looks as though post-allocation demoralisation or similar has caused some of the more content cases in the intervention group (mean score 80) not to bother completing the study, while some of the less content cases in the control group (mean score 65) have been disappointed not to receive the extra help. Whatever the reason, the dropout is not large or unbalanced in number, but it is unbalanced in terms of prior well-being.

**Table 12.3** Initial scores for dropouts in a well-being experiment

|  | Intervention group | Control group | Overall known figure |
| --- | --- | --- | --- |
| Well-being score | 80 | 65 | 72 |
| Number missing | 20 | 20 | 40 |

The 40 cases that have dropped out are effectively not part of the study any longer and their prior scores need to be discounted. If we remove 20 cases with a mean score of 80 from the intervention group, the remaining 180 have a mean of 72. If we remove 20 cases with a mean score of 65 from the control group, then the control group started with a mean of 73. The dropout has now reversed the original position. At the outset, looking at only the complete cases in the control group, they had slightly higher well-being than the intervention group, by 1 point or an effect size of –0.04, even though the true result is that the treatment group were ahead at the start (effect size of +0.08). The figure of –0.04 would be the one the researcher achieved, and is the figure that they would take into account when considering the final outcomes. This could easily lead to an incorrect interpretation of the impact of the experiment.

It seems to the researcher as though the control were ahead at the start, and so if both groups got the same results after the intervention, then it would appear that the treatment group made more progress. In fact, the treatment group were ahead at the start, and so if both groups got the same results later, then it would mean the treatment group actually made less progress. Ignoring the dropout or treating it as a minor issue can lead to the wrong substantive answer. All missing data matters.

In this example we could prevent a serious error in our analysis simply by looking at what is known about the missing cases (Table 12.3). This is, or should be, a standard step in any analysis. If it is not there in any report your read, wonder why.

Another common situation is where you know something useful about missing data but cannot use this to work out the exact scale of the problem. Imagine a survey asking people for their occupation and their income. Table 12.4 shows an illustration where not all participants provide a response for their income, but they do answer a different question about their occupation. Given that we know their occupational group we can at least assess whether the propensity to respond about income varies by occupation. Here, it is clear that the respondents in professional/managerial occupations are somewhat less likely to provide an income figure. Knowing this does not solve the problem, but it can provide a caution about the substantive results about income. The results for income will probably be an underestimate.

**Table 12.4** Percentage of two occupational groups not reporting their income in a survey

|  | Missing an income value | With a known income value |
|---|---|---|
| Professional/managerial occupation | 17 | 83 |
| Routine occupation | 8 | 92 |

Remember that all missing data creates the potential for bias, but this simple approach of looking at what we do know about missing values helps to illuminate the level of bias, and helps us to be cautious and appropriately humble about the strength

## MISSING DATA IN AN EVALUATION OF SWITCH-ON: A REAL EXAMPLE

In an example of a real trial, there were 314 cases (children) randomised to two groups. One group received a reading intervention intended to improve primary school reading, and the other did not (until a term later, in a wait-list design). The research question was whether the intervention led to improved reading scores on a standard test. This study is discussed in more detail in Chapters 5 and 16, and is reported in Gorard et al. (2015). Here we are concerned only with the six cases with missing data, and how this was reported in this real study.

All participants took a reading test at the start and the end of the trial. The average pre-intervention test score for all participants was 76. By the final analysis six children had missing scores for various reasons. One took the pre-intervention test (repeatedly) but his school was unable to record the score properly. Five others took the pre-intervention test but did not sit the post-intervention test (Table 12.5). The missing cases have initial scores close to the average scores for all cases. Although this loss of data, and the reduction of the sample to 308 pupils, are unfortunate, there is no specific reason to believe that this dropout was especially biased compared to the achieved data, or that it favoured one group over the other. Here, the 2% overall attrition probably did not unduly affect the substantive results.

Table 12.5  Pupils allocated to groups but with no gain score, and reason for omission

| Allocation of pupil | Pre-intervention score | Post-intervention score | Reason |
|---|---|---|---|
| Treatment group | 78 | - | Left school, not traced |
| Treatment group | 73 | - | Long-term sick during post-intervention |
| Control | 74 | - | Left school, new school would not test |
| Control | 75 | - | Withdrawn, personal reasons |
| Control | - | 70 | Pre-intervention test not recorded |
| Control | 73 | - | Permanently excluded by school |

## SENSITIVITY ANALYSES

As illustrated so far, it is possible to tell something, often something very useful, about the nature of missing values for any variable when at least some information is known about the cases with the missing values. It is harder to say anything very much about cases missing entirely. Sometimes, of course, the cases will have been selected on the basis of some of their known characteristics, in which situation that information can be analysed in the same way as above.

Otherwise the best approach is a kind of sensitivity analysis, envisaging the most or least favourable substantive findings, by imagining favourable and unfavourable replacement values for cases in the study (Sterne et al., 2009). Sensitivity analyses are used in many fields to help decide whether a finding is robust (in terms of missing data) and so is worth taking notice of substantively (Pannell, 1997; Thabane et al., 2013). The process simply involves looking at the impact on the substantive research findings of varying assumptions about any errors and related factors. One approach is to assess the proportion of cases that would have to be replaced with counterfactual data in order to invalidate the inference being made (Frank et al., 2013).

You could calculate how many of the missing cases would have had to respond quite differently to the results you obtained, in order to change the substantive findings of the research. This is a tough but fair test that soon makes clear how inadequate many published samples are (assuming they report the response rate at all).

Here is an illustration. Imagine we ask 100 employers in retail industries, and 100 employers in leisure industries, whether they envisage taking on new employees in the coming year. All 100 employers respond in the retail sector, but only 80 employers respond in the leisure sector (for an overall 90% response rate). In the retail sector, 50 employers say that can foresee taking on new employees and 50 do not (50% for each). In the leisure sector, 45 employers envisage taking on new staff (56% of 80), but 35 do not. We are about to conclude that there is therefore slightly more optimism about new jobs in the leisure sector, but before reporting we conduct a sensitivity analysis. How differently would the non-responders have to have answered for our conclusion to be false?

If the 20 non-responders were in the same proportions as the actual study, then 11 (56%) would have reported envisaging new employees and 9 would not. It would only take 6 of these 20 people to respond differently (5 saying yes and 15 saying no) in order to transform the study findings. This would mean only 50% envisaging new employees in the leisure industry as well, since 35 + 9 + 6 = 50. So, despite a high 90% overall response rate, the difference observed in this study would not be very secure (6 out of 200 cases responding inconveniently for our findings is not many). The difference found in the research would probably be meaningless.

Thinking like this is good practice. One thing you should definitely do with non-response is to record it accurately, and report it clearly. You must report the results somewhere, with the actual achieved figures. Then at least your readers can conduct this kind of sensitivity analysis, even if you are unwilling or unable to.

More simply, this kind of sensitivity approach can be used with each variable, and for a single sample. For example, imagine you have collected valid data on income from a sample of 950 cases, but actually contacted 1,000 cases. Your response rate is high for social science research, at 95%. The mean income reported by the achieved sample of 950 is £30,000, with a standard deviation of £10,000. Imagine that the missing 50 cases had an income that was one standard deviation above or below that mean – that is, their income could have been £20,000 or £40,000. This is quite an extreme difference, and so the test is a tough one. But if your judgement about your substantive finding would have been the same even if the missing data had been as clearly different as that, then your finding should feel a little safer. You can relax about the missing data to some extent.

If, however, such a difference changes what you think the finding is, then your finding is less secure (see Chapter 8). Here, if the missing cases had values one standard deviation higher than the mean of the achieved sample, then the true mean of your intended sample would be £30,500 (or (950 × 30,000 + 50 × 40,000)/1,000). If you instead tried subtracting one standard deviation, your estimate of the real sample mean would be £29,500 (or (950 × 30,000 + 50 × 20,000)/1,000). If the difference between £29,500 and £30,500 would make a big difference to what you judge the substantive result of your research to be then you should be more tentative, given the possibility of bias created by missing data. If not, you can feel more secure about your results. Here, £30,000 seems a reasonable estimate that would not be unduly disturbed by inconvenient values for missing data, even if the mean turns out to be not entirely accurate.

This kind of thought experiment can be referred to as calculating the number of counterfactual cases needed to disturb a finding (or NNTD for short). The imaginary counterfactual is the inconvenient score for the missing cases, set at the value of one standard deviation away from the achieved mean. If you are comparing two scores (Chapter 5), and you have missing data, then try this. Select the group with fewest cases (if the groups are equal in size, pick either). Select as the counterfactual the mean score for that group plus or minus the overall standard deviation (for both groups combined). Select plus if the mean of the smaller group is smaller than that for the larger group, and select minus if the mean of the smaller group is larger than that for the larger group. Adding the resulting counterfactual score for one case to the smaller group will now slightly reduce the apparent size of any gap between the means of the two groups. This will also slightly reduce any apparent effect size. The NNTD is the number of such counterfactuals you would need to add to the smaller group before any difference between the means of each group disappeared completely.

Suppose we were comparing the average income for two towns (Table 12.6). We set out to get 1,000 responses from each town, but are missing 50 values from town A. Town A has a mean of £30,000 and town B a mean of £40,000. There appears to be a substantive difference in incomes between the two towns, and we are about to conclude that incomes in town B are higher than those in town A, on average. The effect size is –0.8 (or –10,000/12,564), which seems substantial and worthy of further consideration. Just before we go ahead and make this claim, though, we try the tough test of NNTD. We pick the smaller group (A has only 950 cases), and estimate its counterfactual. The counterfactual could be one overall standard deviation (£12,564) above the mean for town A (£30,000). This would be £42,564.

**Table 12.6** Difference between incomes in two towns, achieved sample

|        | N     | Mean income (£) | Standard deviation (£) |
|--------|-------|-----------------|------------------------|
| Town A | 950   | 30,000          | 10,000                 |
| Town B | 1,000 | 40,000          | 15,000                 |
| Overall| 1,950 | 35,128          | 12,564                 |

If we now add one imaginary case to town A with the counterfactual score of £42,564, we get the figures in Table 12.7. The mean estimated income for town A has gone up a little. Recall that the NNTD is defined as the number of such counterfactual scores that would have to be added to the smaller group until its mean was at least as large as that of the other group (town B) and the effect size was zero. If the NNTD is a very large number then the main finding of a clear difference between the means is considered more secure. This is the situation for this example. If the NNTD is small, the main finding is not considered secure and should be presented very tentatively. We would not want policy, practice, or changes in theory, to be based on weak findings.

**Table 12.7** Difference between incomes in two towns, with a counterfactual case

|        | N     | Mean income (£) | Standard deviation (£) |
|--------|-------|-----------------|------------------------|
| Town A | 951   | 30,013          | -                      |
| Town B | 1,000 | 40,000          | 15,000                 |
| Overall| 1,950 | 35,128          | -                      |

Before feeling worried about conducting this sensitivity analysis, note three important points. Calculating the NNTD is much easier than it appears, judging whether the NNTD is large and small is relatively simple, and the whole process does not involve changing any of the data at all.

The NNTD in a comparison such as Table 12.6 is more easily estimated as the overall effect size multiplied by the number of cases in the smallest group. In this example, the apparent effect size was 0.8, and 0.8 times 950 is 760. Therefore it would take at least 760 counterfactual cases of £42,564 to eliminate the difference in means between the two towns. Put another way, we can interpret an effect size as being the proportion of cases in the smaller group that need to be made more counterfactual (or that need to be added as counterfactual cases) before the effect disappears.

Is 760 a large number here or not? The simplest way to decide this is to compare 760 with the number of missing cases or cases missing data in the overall study. Here there are 50 missing values. This means that, even in the unlikely scenario that *all* of the 50 missing cases had a quite extreme score, inconvenient to our claim of a difference between the towns, the finding cannot be explained by missing data alone. There would still be a much higher average income in town B, so our substantive finding that incomes in town B are larger than those in town A, on average, would still be true. Of course, in that extreme situation the effect size would be a bit smaller, but as explained in Chapter 8, our key findings are not usually about the precise decimal places. They are about whether a trend, pattern or difference has been established in the data. In this example, despite missing 50 cases entirely, the result is robust and should stand.

Finally, note that using the NNTD approach is imaginary and does not alter any of our actual data. So, when we report the data it is clear to readers what we actually found, and what is hypothetical like the NNTD. They could then use their own approaches to handling missing data (as described below, and in Chapter 13) if they wish. For more on the NNTD, read Gorard et al. (2017).

Having logged and reported all that we know about missing data, it is now possible to turn to our substantive analyses in order to answer our actual research questions – such as has income changed over time, is it higher or lower than elsewhere, or do older people earn more? Other than sensitivity analyses, how else we might handle missing data in these calculations?

## HOW NOT TO DEAL WITH MISSING DATA

If the argument in this chapter has been followed so far it must be clear that trying to replace missing cases by asking further different cases to provide data will not reduce bias. If non-response occurs for a reason then such 'oversampling' will not help, and may make any bias worse. If you are conducting a postal survey of 100 people, and 30 do not respond even after a reminder, then you can preserve your sample size by adding another 30 cases. Unfortunately, though, even if you try to replace like with like, you will be left with the possibility that those who do respond are more alike in some way than those who do not. They may be more literate, more likely to have time to respond, and so on. It is better, as above, to be as creative as possible about ways to discourage non-response, and to follow up any non-responders.

A further approach to estimating the bias that has been suggested is to look at the order in which people respond. An argument can be constructed that those who reply later to a postal or online survey, for example, are more like those who do not respond at all than are those who reply early. Thus, it is argued, we can estimate the character of non-respondents by looking at the difference between early and late responders. This is certainly worth a try, and there is no harm in recording when and how a response appears, and how much cajoling/reminding was necessary. However, studies have suggested that the empirical basis for this approach is weak (Giacquinta and Shaw, 2000). Using late returners to estimate the sample bias induced by non-returners may not be very effective, and may even lead to a poorer estimate. There is no clear evidence that an estimate of propensity to respond (based on how many requests need to be made, or how late the response is) can be used to generate valid data similar to the cases that do not respond at all.

## WEIGHTING THE RESPONSES

Some analysts try to adjust for non-response or missing cases in their achieved samples via a process of **weighting** the results they do have. If we know something about the characteristics of the population from which our sample is drawn, and it is clear that the achieved sample differs from the population in terms of such characteristics (perhaps there are too few of one ethnic group), then we could try artificially inflating the role of the substantive scores (e.g. the incomes) for the cases with the underrepresented characteristics (Lehtonen and Pahkinen, 1995). This does not increase the number of actual cases, but might conceivably prevent some of the bias created by those missing cases. If the underrepresented ethnic group has a high mean income, then our achieved result would underestimate overall average income. Weighting is intended to address this.

Imagine collecting data from a sample of 1,000 residents, consisting of 600 (60%) of one ethnic origin (group A), and 400 (40%) of another ethnic origin (group B). One of the other variables is whether residents considered voting for more than one political party at a recent election. The overall result was that 440 (44%) had considered another party, and therefore the modal average (most frequent) response was 'no' (56%). Separated by ethnicity, 60% of the group A respondents but only 20% of the group B ones had considered voting for another party. This is summarised in Table 12.8.

**Table 12.8** Ethnic group by political flexibility, before weighting

|         | Yes | No  | Total |
|---------|-----|-----|-------|
| Group A | 360 | 240 | 600   |
| Group B | 80  | 320 | 400   |
| Total   | 440 | 560 | 1,000 |

The census of population for the region suggests that the proportion of ethnic groups should be 80:20 for group A and B residents (not the 60:40 in our sample). The sample overrepresents group B residents. If the achieved sample had 800 group A and 200 group B cases, and both groups had answered the voting question in the proportions achieved, then 60% of the hypothetical 800 group A cases (480) and 20% of the 200 group B ones (40) would answer 'yes'. On this calculation, since ethnicity makes such a difference, and the sample overrepresents the views of group A residents, the best estimate of the population figure considering another party might be 520 per 1,000, or 52% instead of 44%. The modal average response would actually be 'yes', even though the achieved sample appeared to suggest 'no'. This would be the answer weighted for the proportion of ethnic groups in the population.

Weights like this can only be used post hoc to correct for variables for which true population values are already known (here the ethnic breakdown among the population), often making weighting impossible in practice. And weighting a sample in this way clearly cannot correct for other variables for which the true population value is not known (Peress, 2010).

Weighting also quickly becomes complex in real-life research because many such variables may be involved. The complexity of weighting arises in the interaction of the variables. If, in the example above, you decide that the sex of respondent is another important factor then you will need to consider the four groups consisting of male and group A, female and group A, female and group B, and male and group B, separately. If you add first language, level of education or social class then the calculations become mind-boggling. And these are only some of the standard contextual variables we might want to use.

The dangers of being misled by weighting are probably greater than any gains. Here is an extreme example to make the point. Imagine a large survey of adults with an overall response rate of 99% but in which the response rate from the small Traveller community was only 1%. Would it be reasonable to multiply the results obtained from a few Travellers by 99 to estimate what the majority of missing Travellers would have said, weighted at the same scale as the rest of the respondents? This would be very misleading if the 99% of Travellers who did not respond were different in some ways from the 1% who did. Weighting is not really the answer.

## COMPLETE CASE ANALYSIS

It is quite common for analysts to simply ignore missing data, working with the existing cases, and omitting any cases where there are missing values. This approach of working only with complete cases can create several problems.

Most obviously, if the cases that are missing data for any variables are simply ignored then this reduces the number of valid cases, and so limits the appeal and trustworthiness of the analysis (Sterne et al., 2009). In an analysis that involves dividing the sample into two or more groups for comparison, there may be too few cases in any table cell for a worthwhile result. In a multivariate model such as regression (Chapter 16), ignoring cases with any missing values can reduce the number of valid cases substantially – perhaps even to zero – in a dataset with many variables. This is so, even if the number of cases missing values for each variable is small, because any case with any missing values in any of the variables will be ignored.

A real example is based on using the National Pupil Database (NPD) for England which contains an official record for every student registered at a state-funded school (but which is available to researchers). The NPD has hundreds of variables, and most of these are completed by schools, as defined and required by law. The NPD is one of the most complete datasets possible. Even so, many key variables might have up to 15% missing values (such as where the ethnicity of a student is not declared). A relatively simple analysis based on as few as six variables, and that excluded any case with missing data, could easily find itself facing 50% or more loss of data in total (e.g. Gorard, 2010c).

Second, data based only on complete responses is almost certainly biased, as can be demonstrated when population or administrative data is compared to incomplete surveys of the same cases, or where different studies have selected exactly the same 'sample', but then have different response rates – useful examples appear in Dolton et al. (2000) and Behaghel et al. (2009). Unless the data that is missing is missing completely randomly (a very unlikely scenario), ignoring the whole case where any values are missing will bias the data and so produce knowingly biased results (Swalin, 2018). This should be avoided.

In a longitudinal study, if the poorest respondents are more likely to drop out from later waves of data collection, then estimates of average household income will become seriously inflated (Siddiqui et al., 2019). In secondary data, the most disadvantaged cases often do not have valid data even about their level of disadvantage (Gorard, 2012). This then distorts both the policy and practice implications from any analysis using that data. For example, schools in England are meant to receive additional funding for every disadvantaged student in their intake. If the disadvantaged status of a student is not known, the school will not receive the funding but will still have to deal with the challenges that this student might face.

In an RCT with two groups, if the more proficient cases in the treatment group find the intervention dull and are more likely to drop out, and the less proficient cases in the control group are disappointed not to be selected for the extra help, and are more likely to drop out, then the results will be distorted. The study may suggest that the intervention was ineffective, regardless of its actual benefit, because the post-intervention scores for the control group will be artificially high. Examples like these can occur in any real-life study.

Third, when research-based measurements are used in calculations, any initial errors in those numbers then propagate through the calculation (Chapter 14). For example, subtracting two numbers of similar size, each with errors caused by missing data or anything else, can yield a small answer that is almost entirely composed of the error parts from the original numbers. This can distort the results of an evaluation such as an RCT (Gorard, 2013a), and is a major problem in studies of school effectiveness or similar (Gorard, 2010c).

Let us return to the example based on comparing the average income of two towns. Imagine that our estimate of the income for the first town is £21,000, but it should be (unknown to us) £30,000. The difference is largely caused by missing data bias. The error in our measurement of £9,000 is 30% of the real score of £30,000. Our estimate for the second town is £30,000, but this is also an underestimate. The true mean income is £32,000. The error in our measurement of £2,000 is 6% of the real score of £32,000. The actual difference in mean income between the two towns is £2,000 (32,000 – 30,000). Our research would suggest that the difference was £9,000 (30,000 – 21,000). Our estimate is out by £7,000, and this is now 350% of the size of the real difference (7,000/2,000). Our initial measurement bias of 6% and 30% has propagated to a massive 350% error in the results, just by simple subtraction. Our research result is therefore almost entirely made up of error. This kind of situation probably occurs in most research, and will be almost universal in the very common small-scale studies in social science that take no account of bias or missing data.

There are many good reasons not to simply ignore missing data by conducting a complete case analysis.

A further reason that is sometimes given for dealing with missing data is that otherwise the resulting bias will affect any standard errors estimated, and so any significance tests, confidence intervals, and other results based on these estimated standard errors (Chapter 7). This reason is not a good one, and does not make sense. The use of significance tests and related approaches is not relevant unless the cases are fully randomised. And datasets with missing cases and values cannot be random, *by definition*. Nor can a dataset with missing data be made random subsequently. Therefore, even if they worked, significance tests cannot be used if there is missing data. However sophisticated any approach is, this whole idea is an illusion.

## COMPLETE VALUES ANALYSIS

The previous section outlined the notion of complete case analysis, wherein all cases with missing values are dropped and computations take place only with the remaining full cases. An alternative is to keep all cases, and drop only those that have missing values for the variables to be used in each analysis. So, if variables A, B and C all have missing values for different cases, then a comparison of the complete cases for A and B would include different cases than a comparison for A and C, etc. Both would involve more cases than a multivariate analysis of A, B and C, and more than a bivariate analysis based on complete cases. Despite retaining more cases for some analyses than complete case analysis, this approach is even more dangerous. The number of cases obviously varies for each analysis, and so the precise subset of cases changes as well. This can be very misleading, for example, if you wanted to compare the correlation coefficients for A and B, and A and C.

## REPLACING WITH DEFAULT VALUES

Other than sensitivity analyses, nothing very much can be done about completely missing cases. What else could be tried to keep cases with missing values?

A simple approach would be to replace any missing values with a default value that allows all cases to be included in the substantive analysis, without that replacement value having an undue influence on the findings. A reasonable default value for any real number variable (that is not a dependent variable in a regression model) could be its overall mean score for the cases with values. This preserves all cases, but the default value does not change the mean for that variable, and the cases with the default have no extreme leverage on the substantive findings. If this is done, it is important also to create a new variable representing whether the original value was missing or not. This allows analysis of the known characteristics of missing cases (as above), and also allows 'missingness' for that variable to be used as an additional explanation in any substantive analysis, if you want.

An equally simple approach for categorical variables is to create a new category of 'missing', and use this for analysis of the known characteristics of missing cases, and as a possible explanation in any substantive analysis. For example, a binary variable recording whether a respondent had attended university or not might be recoded as having three possible outcomes – yes, no and not known.

Another approach for categorical variables is to recode the missing values as one of the existing categories. For example, a binary variable recording whether a respondent had attended university or not might be recoded as whether a respondent is known to have attended university or not by converting missing values to not known to have attended (the coding becomes yes or not yes, rather than yes or no). As with the use of means for real numbers, and for the same reasons, a new variable should then be created recording whether the original value was missing or not.

These simple approaches mean that all cases can be used, while allowing further analyses of missingness, and including missing as a predictive variable in the substantive analysis. Used with otherwise high-quality datasets, such techniques can lead to somewhat more powerful substantive models than analyses just using complete cases (e.g. Gorard and Siddiqui, 2019).

However, these approaches will tend to reduce the apparent variability (e.g. standard deviation) of the sample for any variable involved, because more cases will now have the same value/category. If this is seen as a problem (by making effect sizes misleadingly large, for example), the replacement values could be adjusted by adding/subtracting a small random component to/from each.

Alternatively, and more simply, an analyst could use the standard deviation of only the complete cases when computing effect sizes (another reason for keeping a record of all initially missing values). This is simple and safe. It is also better to insist that, in a large dataset, only a small fraction of values be replaced for any variable using this method. If a variable has a very high proportion of genuinely missing values then it is better to treat the whole variable as non-viable and discard it. Then, using the record of how much missing data there was originally acts as a caution to both the researcher and reader against overinterpreting small 'effect' sizes. This is also where a simple sensitivity analysis (see above) is valuable.

### Exercise 12.8

Why do you think that researchers who do not work with numbers do not generally consider missing data or the bias it creates? Is this justified?

## CONCLUSION

In social science, it is still the case that too many research reports ignore the issue of missing data, not reporting it at all, or not reporting clearly how missing data has been handled (Berchtold, 2019). Most authors do not publish even basic information about their response rates, where these are relevant. Where research reports do cover missing data, it is not clear that missing data has always been handled appropriately. In most real-life datasets there will be missing cases and missing data, and for most of these the data will not be missing randomly. The missing data creates the likelihood of bias in the achieved results.

It makes sense to report all possible forms of missing data, report everything that is known about the characteristics of missing cases and of cases missing values, and conduct basic sensitivity analyses of the potential impact of these missing data on the substantive results. Simple default replacement values can be used, to retain all cases for analysis, as long as knowledge of which cases are missing data is retained and used for further investigation. And as long as this affects only a minority of values for any variable.

Bias in the substantive results caused by missing data generally cannot be corrected by any more technical means (Cuddeback et al., 2004). Missing data cannot be accurately replaced by using the evidence that *has* been collected. In fact, attempts at such replacement often make the bias worse.

The next chapter is for those who want to know more about missing data, and why the more complex approaches that you may read about are not needed.

### Notes on selected exercises

#### Exercise 12.4

Many issues relevant to missing data can arise when comparing two or more sets of data. In general, the more variables and groups there are, the more problems there will be with missing data. A common problem that goes largely unremarked is that comparing two values, such as the average income of the residents in two towns, creates a much larger error component in the answer than in either of the original figures. The average income of the two towns will be of similar magnitude, even where there is a noticeable difference between them. Each will only be an estimate of the true average income for that town, because of errors in data collection and measurement, and because of missing data. The maximum scale of the error in each average is added when the two values are subtracted (because we do not know what each error is, but it could be positive or negative). This larger error component is now the error in a much smaller figure (the difference between the averages). The maximum error relative to the answer will grow hugely. The more complex the calculation is, the worse this problem can become.

## Exercise 12.5

Academic research reports are often made needlessly hard to read, as though the author has no care for the reader. Whether the study uses numbers or not, the abstract should contain a statement of the problem or question, why this might matter, how the issue was addressed, what results were found, and what these results might mean in terms of the problem or question. These steps can then be explained in detail in the full report. If the author is not clear, this generally means that they are incompetent (careless) or that they are trying to hide something from the reader. Be explicit as an author and demand the same from others.

## Exercise 12.6

a   The response rate, as far as we can tell, is 500/1200 or 42%. This is quite poor, but higher than in quite a lot of published studies. It means that we can really tell very little about the population. It also means of course that we cannot use any sampling theory statistics, such as significance tests, since these assume full response.

b   In the achieved sample there are more middle-class families (300/500). If all cases approached (1,200) had taken part, the minimum number of families needed for there to be a majority of working-class families would have been 601. This means that if 401 or more of the 700 non-responding families were working class then working class would (and should) have been a majority (along with 200 already classified as working class). So the study has to argue that there is no reason to suppose that the missing cases were this much more likely (401/700, or 57%) to be working class. If, on the other hand, we can think of a reason why working-class families would be less likely to take part in a household interview then we might want to reject the study findings (in this respect). Note that this is not a technical issue, nor even a statistical one, in the sense in which the term is usually intended.

c   This gets a bit complex to envisage even though there are only two variables to consider (class and car ownership). This difficulty of conceptualisation may be part of the reason why error theory and error propagation are largely ignored in other methods resources (in favour of easier but fake solutions such as significance testing). A 10% misclassification rate is perfectly realistic when converting interview responses into a rather subjective category system such as occupational class. There are 300 middle-class and 200 working-class respondents. If 10% of each are in the wrong category, then 30 apparently middle-class families are working class and 20 apparently working-class families are really middle class. Thus, the research should really have

*(Continued)*

identified 290 middle-class and 210 working-class families. This 10% misclassification alone does not substantially affect the proportions of each class. A 10% initial error leads to a relatively small error in the result. This is acceptable, and quite normal both in social science and beyond. However, the situation is not as simple as that.

We do not know how many car owners were in each of the misclassified groups. Originally, it seemed that 66% of the middle-class (200/300) and only 50% of the working-class (100/200) respondents owned cars. This seems a sound basis to declare a difference. But if all of the misclassified middle-class (30) ones owned cars and none of the misclassified working-class ones did (20), then the apparent difference in car ownership disappears. There are now 170/290 or 63% car owners in the middle class, and 130/210 or 62% in the working class. The apparently substantial difference has disappeared at a stroke! This may seem extreme, but the principle of looking at how safe any results are is a good one, if there are errors in the initial data. Of course, it gets a lot more complicated in most real studies. Even in this simplified example, we have ignored so far the possibility that there is also some misclassification in reporting car ownership as well. Imagine building that into the calculation, and then imagine how to track the propagation of errors and their possible implications in a multivariate analysis. Simple is good, in design, in logic and in calculations.

## Exercise 12.7

The obvious danger is that by seeking convenience for your sample and response rate, you are ignoring the data you might collect from a potentially important subset of the population. If you were doing a survey about experiences of injustice in schools would you ignore those who could not read/write or with visual impairment? If you were interviewing people about their experiences of migration would you ignore those who could not speak the language of the interviewer? Hopefully not in either example. Seeking convenience can distort research.

## Exercise 12.8

It is not at all clear why researchers who never work with numbers do not generally consider missing data or bias. They generally do not go through the standard stages of sampling, do not report non-response, and do not report the limitations of their studies, or estimates of bias in their samples. But this is far from the biggest problem. Most give no consideration at all to study design, and they then work with samples that are far too small for them to make sensible and robust internal comparisons between subsample groups. Their work is generally more like journalism than research, and defended in terms of purported paradigms and methods identities.

---

## Further exercises to try

Pick a research report or paper you are interested in, one that uses numeric analysis of any sort. Look at how it describes the number of cases. Is the number of cases clearly stated, or obvious by implication, for each analysis presented? Does it have a CONSORT flowchart or similar? How are missing cases handled? Are you happy with how missing data has been dealt with?

## Suggestions for further reading

An example of the practical importance of missing data in a real study:
Gorard, S. (2008) Who is missing from higher education? *Cambridge Journal of Education*, 38(3), 421-437.

An example of how missing data can be used in substantive analyses:
Gorard, S. and Siddiqui, N. (2019) How trajectories of disadvantage help explain school attainment, *SAGE Open*, 9(1). https://journals.sagepub.com/doi/10.1177/2158244018825171

A whole book on missing data, for those who want to pursue interest in this:
Allison, P. (2001) *Missing Data*. London: Sage.

# 13

# HANDLING MISSING DATA: MORE COMPLEX ISSUES

### SUMMARY

Chapter 12 looked at how missing data might arise in any dataset, how it could be recorded and analysed, and how it might be replaced or handled safely. This helps retain as many cases as possible without distorting your findings. Using 'missing' as a valid value in your analyses, and replacing missing values with a default value (with care), is a perfectly appropriate way of handling and respecting missing data. And conducting simple sensitivity analyses is always recommended, whatever other approaches you use. This is really all that you should need to know.

However, you will come across more complex approaches, and terms like 'missing at random' and 'multiple imputation' in your reading of some research, and in some methods resources. So, this more technical chapter about missing data explains what these terms mean, why you do not need to use these ideas for your own studies, and why the approaches involved might even harm your findings. Not using them will certainly make your research easier to read and understand.

### MODELLING SIMPLE IMPUTATION

When commentators refer to **imputation** as a way of replacing missing values, they generally mean something more than using default values of the kind in Chapter 12. You could, for example, try to work out what any missing value might be by considering the values for that variable in the cases that do have a value. You might be lucky enough to find one or more complete cases that are identical, in terms of *all* other variables, to a case that is missing one value. If so, you might use the values you do have to replace the value that you do not have.

A simple example appears in Table 13.1. Imagine that you have done a survey, and collected three variables from a large number of cases. One case did not provide data on their monthly rent (fourth column of Table 13.1). But happily there are two other cases that produced the same results for both other variables of year of birth and ethnicity (second and third columns). If we find the average rent for these two cases, this could provide an estimate of the missing rent for case 3. This estimate would be 510 (or (550 + 470)/2). We might therefore replace the missing value for case 3 with 510, and proceed to our substantive analyses with (apparently) complete data for all cases.

**Table 13.1** Two complete cases and one case missing one value

| Case number | Year of birth | Ethnicity | Monthly rent |
|---|---|---|---|
| 1 | 1987 | Chinese | 550 |
| 2 | 1987 | Chinese | 470 |
| 3 | 1987 | Chinese | missing |

### Exercise 13.1

How valid does this procedure appear to you? Discuss the limitations and dangers of the procedure, perhaps in relation to the simpler idea of treating 'missing' as a valid result in itself.

Of course, this example is unrealistic in several ways. We would normally collect many more than three variables in any survey, and this makes it less likely that all variables would match each other in more than one case. And there would be more than one case missing any value, and there would be data missing from most variables. Again, this makes it more unlikely that we find cases with exactly matching values for all variables, except for just one which is missing. Much more importantly, the procedure still does not address the problem that missing data is likely to be missing for a reason. If, in the example, the cases missing a value for monthly rent were all homeless (with a monthly rent of zero), then by replacing their missing values with the average rent for any number of non-homeless cases we would be seriously biasing our overall estimates of rents. It might be better simply to report the average rent for all cases that responded to this question, rather than pretending that we have a true average that includes homeless respondents as well.

Another concern that has been raised is that if more than one case is missing a key value, and these cases are alike in all other variables, then this procedure would replace the missing data for all such cases with the same value. The same issue arises if we replace missing data with the overall mean of non-missing cases (Chapter 12).

Both approaches will tend to reduce the apparent variation of the responses, giving a smaller standard deviation than the true standard deviation would be if there were no missing values (for a real number).

One way of tackling this issue would be to add/subtract a small random number to/from each replacement value, so that all are similar but slightly different. In Table 13.1 we might replace the missing value with 504, 510 or 513, depending on the small random number we generate. We would have to decide on the range of random numbers involved, and this should be proportionate to the size of the mean for the variable that has missing values. The range could be related to the standard deviation for the values we do have (we cannot use the true standard deviation because we have missing values, of course!). We would then end up replacing the values we do not know with the biased values we do know, and then modifying them using a number also created from the biased values we do know. It is not clear that this necessarily reduces any bias created by the existence of data that we do not know.

At the next level of complexity, imputation is done by creating a predictive regression model (Chapter 17). All of the variables in the dataset, except one, are used to create an imputation model that predicts the missing values for that one other variable. The dependent or predicted variable is the one with missing values, but the cases with missing values for that variable are ignored for the present. The independent or **predictor** variables are all other variables that might be relevant. The model uses the predictor variables to predict or explain the existing values in the variable with missing values (but ignoring the missing values). The best prediction is then used to create the possible replacement values. In summary, the replacement for any missing value is the multivariate composite value that similar cases have, who are not missing a value for this variable.

This is a more complicated way of doing what was tried with Table 13.1. It makes the dataset complete in terms of that variable, and so the substantive research analysis can be conducted with a full dataset. As above, a random element can then be added artificially to create more variation in the replacement values. This approach has the same limitations as the simpler method, and these are discussed further below.

## MULTIPLE IMPUTATION

Multiple imputation is the most complex approach to handling missing data that is relatively widespread. It is, in effect, the same modelled imputation as described above. However, this procedure is repeated many times (hence 'multiple'), creating many full datasets by estimating different plausible replacement values each time, because of the random element in the model approach. This adds variability to the predicted missing values representing the proposed distribution of the missing data, given the data that is not missing (Brunton-Smith et al., 2014). Each complete dataset is then used to run the full substantive research analysis, and the multiple results are combined (or averaged) to give an overall result and an estimate of the variation within that. In reality the process is more complex than this, because it is rare for a large dataset to have only one variable with missing values. Therefore plausible values are needed for more than one variable at a time. This sounds, and is, complex.

Other than its complexity, multiple imputation has several disadvantages for the analyst. It is not clear that even if the missing values are strongly patterned in terms of other available variables, their replacement values should be the same as or very similar to the values for cases not missing data (as with simple imputation above). Like simple imputation and replacement, multiple imputation is completely unable to handle missing cases (it only deals with missing values for existing cases). So, even if it worked and was justified, a major source of bias remains. The same is true of other replacement approaches, but with multiple imputation in particular this fact is often ignored as though this complicated procedure was some kind of panacea. However, the main reason why neither simple nor multiple imputation should be used is that both rely on an assumption that all missing data is 'missing at random'. What does this mean?

## THE THEORETICAL NATURE OF MISSING DATA

So far, we have considered missing data as being a cause of possible bias in our datasets, as opposed to the less practical idea that missing data occurs by chance. Clearly, some missing data could be due to chance (occurring randomly), but it would be a mistake to assume that all missing data arises for no reason (rather than being linked to lower literacy or motivation, mobility or homelessness, for example). However, many 'authorities' writing about the conduct of research are not happy with this conclusion.

Instead there is an idea promoted in social science that there are three generic types, or causes, of missing data (Little and Rubin, 2002). The first clear type is where the loss of data has occurred randomly (see Chapter 10), and so the missing data (if it could be found) would be an unbiased subset of the full dataset – both the missing and non-missing. There would be no patterning of the missing data in terms of any of the other variables. And the reason for any data being missing would be unrelated to its true value. This is a theoretical description only. It is hard to envisage this happening for any real-life social science dataset, and it is generally impossible to identify such a situation in practice using only the data that is not missing. This is because, if we had access to the data that is purportedly missing, then it is not missing. We do know that non-responders to a survey may tend to be busier, or less literate, confident or interested in the topic. This is common in social science, and it is definitely not a random phenomenon.

The second clear type is where the missing data has not occurred randomly, and whether it is missing or not depends partly on, or is linked to, the missing values themselves (as in Chapter 12). For example, the average level of literacy of respondents might be overestimated, where less literate cases are missing from a survey dataset, because these respondents are less likely to be able to read a questionnaire. Or the estimated literacy level may be biased because of another variable in the dataset. For example, the average level of literacy may be overestimated where homeless cases are somewhat more likely to have low literacy, and did not receive a postal questionnaire because they have no address to post to. Both kinds of biases are possible wherever data is missing for any reason, and these biases have been found to be very common in those rare situations where it is possible to judge the patterns in the missing data by comparing them with a more complete dataset (Dolton et al., 2000). This is the kind of problem addressed satisfactorily in Chapter 12.

The third type of supposed missing data is much less clear, and is rather confusingly referred to as 'missing at random', as opposed to the first type which is referred to by some writers as 'missing completely at random', and the second which is 'missing not at random'. Here the missing data is assumed to be missing for reasons unrelated to its value, but only after controlling for some other variables. The theory is that the observed data can be used to explain or predict the missing data.

Put another way, it means that any pattern in the missing data can be fully explained by looking at the values of other variables in the dataset (Brunton-Smith et al., 2014). For example, a local authority might have records of the attainment of students at school in its area. Some students aged 16 may have moved elsewhere, and their attainment at age 16 is missing from the records. Looking at the prior attainment of the missing cases at age 11, it may be clear that the average of the missing attainment scores at age 16 will tend to be lower than the average of the cases for whom data is available. Therefore, the missing data is conceived as not truly random. However, the 'missing at random' idea is that if we control for attainment at age 11, which is strongly correlated with attainment at age 16, then attainment at age 11 explains the association between attainment at age 16 *and* the likelihood of missing data for attainment at age 16. Once controlled for, the data is said to be 'missing at random' (but is still not actually missing randomly, as in the first type of missing data).

As another example, if we assume that the missing cases in a survey are due to the cases being homeless, but we have data for some homeless cases, then we might use the average literacy of the homeless cases we do have, to adjust for the literacy of all missing cases. This is a very unrealistic scenario, and it can never be tested for accuracy because we do not know the values of the missing data, and so cannot compare cases with and without that value, to check whether they differ in terms of that variable (Soley-Bori, 2013). It is very likely that the homeless cases that did respond to the survey were different from those who did not in a way that would affect the literacy estimate. For example, if the urban homeless were more likely to respond and had higher literacy, on average, than rural homeless cases then using the figures we do have in order to estimate the figures we do not have could be ineffective or could even create more bias. The assumptions underlying 'missingness at random' are unrealistic. For example, it would be hard to be confident that all missing data on literacy was due to homelessness (as opposed to illness, motivation or being on holiday). And it is unwarranted to assume that any homeless people who did respond had the average literacy rate for all homeless people. The problems here are like those of weighting.

**Image 13.1  Neglected by research?**

Also, in a large dataset where some data is missing from one variable there is also likely to be missing data for other variables, making it unclear which missing values should be addressed first. In the example above, some cases might be missing figures for their attainment at age 11, and others might be missing figures on their attainment at age 16. If those missing data from age 11 are lower attaining, on average, then using the existing age 11 figures to control for missing age 16 attainment figures might create more bias than leaving well alone. We could try and control for the missing age 11 figures as well, but this would then require a third strongly correlated variable, which might also have missing values, requiring a fourth variable which would also have missing values, and so on.

Values 'missing at random', as defined by Little and Rubin (2002), are meant to be predictable, patterned by population characteristics, and can be summarised (or modelled by regression) more efficiently than by simply listing each one. Therefore they are not random, by any of the philosophical/scientific definitions of randomness in

Chapter 11. And none of the considered definitions of randomness allow for degrees of randomness – it is clearly something that either is or is not.

The biggest limitation of multiple imputation, therefore, is that its advocates say it should not be used where missing data is not random (the majority of real-life situations) or where the data is missing completely randomly (presumably the rest of any real-life situations). And if the missing data is not actually 'missing at random' (the supposed middle ground between these two), then using multiple imputation may well lead to greater bias (Hughes et al., 2019). However, it is not clear that there *is* or can be any middle ground – where data is missing just a bit randomly – and so it is not clear that multiple imputation should ever be used.

The identification of this missing at random, but not actually missing randomly, status (the middle ground) cannot be determined by examining the data that does exist, and instead must be merely assumed by any analyst (Crameri et al., 2015). 'Missing at random' is an assumption that is used to justify the analysis, not a property of the data being examined. Sterne et al. (2009) use the example of a study of the predictors of depression. If people with depression are more likely to drop out of a study, or miss an appointment, because they are depressed, then the necessary missing at random assumption is not viable. All of the examples in the book so far, such as longitudinal dropout patterned by household income or homelessness, would be unsuitable for multiple imputation for the same reason.

The main argument usually marshalled for assuming 'missing at random', and so being able to use multiple imputation, is that any other approach can cause the associated standard errors in the results to be estimated as smaller, because other approaches do not take uncertainty into account sufficiently (Sterne et al., 2009). This would cause significance tests to produce even more spurious 'significant' results. A contradictory claim is that if we do not use multiple imputation, then there is an increased chance of missing a significant result (van Buuren, 2018). But neither issue matters at all if the substantive analysis that follows does not use significance tests, as it should not (Chapter 6). Therefore, the main argument for using multiple imputation no longer exists.

Multiple imputation does not automatically yield safer results than the simpler approaches in Chapter 12 (Soley-Bori, 2013; Swalin, 2018). It is also noteworthy that under many common conditions multiple imputation gives the same substantive results as complete case analysis anyway (such as when the missing values are in the outcome variable, or when conducting logistic regression analysis, described in Chapter 18). In these circumstances, all of the complexity faced by researchers and their readers is for no purpose.

## CONCLUSION

Analysts attempting to deal with missing data do so for a number of reasons. They may want to retain as many cases as possible, or to create substantive results that are as unbiased as possible (or at least not to be misled by bias). However, as this chapter suggests, these aims will not generally be achieved using complex approaches, chiefly because the underlying assumptions of these approaches are so rarely met in real life. There is considerable doubt that bias in substantive results caused by missing data can ever be corrected by technical means (Cuddeback et al., 2004). And neither single nor multiple imputation can do anything about the potential for bias caused by non-response and missing cases. Using these needlessly complex techniques also makes research reports harder to follow, and so harder to check for trustworthiness.

In summary, it is safest to assume that data is seldom missing truly randomly, and that there will always be a pattern created by the reason that at least some of the values are missing. All missing data therefore creates a possibility of bias in any results based on the remaining cases (Behaghel et al., 2009; Peress, 2010). It must be dealt with accordingly, by retaining the notion of missingness in your analyses somehow.

## Suggestions for further reading

More on the ideas in this chapter:
Gorard, S. (2020) Handling missing data in numeric analyses, *International Journal of Social Research Methods*. https://www.tandfonline.com/doi/full/10.1080/13645579.2020.1729974

A journal article on missing data for those who want to pursue the topic:
Dong, Y. and Peng, C. (2013) *Principled Missing Data Methods for Researchers*. Springer Open. https://springerplus.springeropen.com/articles/10.1186/2193-1801-2-222

# 14

## ERRORS IN MEASUREMENTS

### SUMMARY

This chapter describes some of the major sources of errors in measurements. These issues are relevant to how much we trust the kinds of numbers used in social science research, or how 'valid' they are judged to be. It also looks at the issue of the reliability of repeated measurements, such as those often used in attitude testing. This chapter is a bit more conceptual than most, and can be skipped by a reader who wants to press on to issues of correlation analysis and statistical modelling.

### ERRORS IN MEASUREMENTS

An obvious limitation for any number measurement scale is it proneness to error. All measurements are prone to error, and many measures used in social science contain a very high proportion of error. Think of attitude scales, the allocation of national examination grades, or levels of poverty. What they all have in common is a high level of imprecision – some, like attitude scales, because of the vagueness of what is being measured, and some, like the allocation of national examination grades, because of the size and difficulty of the operation lead to mistakes and imperfect moderation between assessors, years and different subject areas.

For any measurement, you may be missing at least part of the numeric information you need (see Chapter 12). Or there may be a difference between the measurement you should assign to the reality using your chosen scale, and the measure you do assign. This is a simple mistake of the kind that we all make, especially when conducting repetitive tasks. One common source of simple mistakes comes from copying numbers, such as when transferring a list of numbers from paper to a computer or calculator. While we can take steps to minimise the chance of such errors, we cannot guarantee their absence.

Key components of measurement are knowing whether we have the correct figure, and how far out our measurement is from the true figure. A measure is intended to react in the same way as the quantity it is supposed to be measuring (Gorard, 2003a). Measurement error is an indication of the failure of that isomorphism, where the measurement does not accurately portray what it is intended to. This kind of measurement error has nothing to do with sampling variation. Measurement error can come from misspecification of the scale, miscalibration, misreading, misrecording, and errors in transcription or recall.

We have no reason to assume that such sources of error are random in nature. In fact, there is a considerable body of evidence showing that they are non-random. Researchers are more likely to misread or mistype data in such a way as to support their favoured ideas, for example. Similarly, a ruler that was calibrated to be too short would consistently overestimate lengths, rather than provide any kind of random variation around the 'true' length. There may be random errors as well in operating this ruler, or in recording the scores generated from it. But these would be additional to the underlying bias, which would remain even after a large number of similar measurements had been aggregated. Thus, the techniques of statistical analysis based on random sampling theory (Chapter 6) are of no use in assessing the quality of a measurement.

In many areas of social science we cannot even begin to estimate how large this discrepancy is since we have nothing to calibrate our 'measures' with (Gorard, 2010a). For example, if we use a proprietary test to assess the mental well-being of someone, because we cannot tell this just by observation perhaps, then how can we assess the measurement error? We may only have the test result. This also means that we cannot be sure that small and even medium-sized differences between many such measurements are meaningful, rather than just the result of error.

There is no standard acceptable amount of error in any measurement. The relevance of the **error component** is a function of the scale, and of the use to which the measurement is put. However, the size of the error relative to any result is not determined solely by the accuracy of the original measurements. It depends on the precise steps in the ensuing analysis. Of course, it helps if the initial readings are as accurate as possible, but whether they are accurate *enough* depends on what is done with the readings, and how the error component propagates as it is used in calculations. It is important to recall that every statistical, arithmetic or mathematical operation conducted with measurements is also conducted with their error components. If we square a variable, then we also square its error component and so on. The more complex the calculations that we conduct, the harder it is to track the propagation of measurement errors, and so make an informed judgement of the ratio of error to the final result.

In extreme cases, the manipulation of variables leads to results almost entirely determined by the initial errors, and very little influenced by the initial measurements.

When answering a typical analytic question, such as whether there is an important difference between two figures (Chapters 5 and 8), we will need to take into account, in a way that traditional methods resources generally ignore, the likely scale of any errors.

## ERROR PROPAGATION

The previous part of the chapter introduced the idea of errors in social science data, how we must assume that all data has some error component, and how **error propagation** occurs when data is used for analysis. This latter issue is important for analysts to recall. It has no technical solution, emphasises the need for good strong measures, and accuracy, and explains why differences or patterns in data can be entirely meaningless. It should keep analysts humble. Regardless of how the initial errors creep into our data, from missing cases, missing data, discrepancies in measurement, and errors in data entry or storage, it must be assumed that any resulting dataset will contain errors. None of these sources of error is random, and none can be dealt with by using sampling theory and its derivatives.

If we have two measurements with their initial measurement errors, and subtract one from the other, then the answer is also in error by some amount. Since we do not know whether either of the original measurements was too large or too small (positively or negatively in error) we must assume the worst. The worst case is that one measurement was too large and one too small, in a way that means our subtracted result includes the sum of these errors, while the subtraction has reduced the size of the apparent answer. This creates a maximum relative error in our result that is considerably larger than in the initial measurements.

This is clear if we imagine that our initial measurements were 1,000 and 990, and that both were only 90% accurate, having a maximum relative error of 10% each. This means that the first number could actually lie between 900 and 1,100. The second could be anywhere from 891 to 1,089. When we subtract our two measures we get 10, but the true answer might be anywhere from −189 (900 − 1,089) to +209 (1,100 − 891). Put as a relative error, our computed answer of 10 could be as much as 199 away from the true answer, so the maximum relative error in the result is 199/10 or 1,990%. The simple act of subtraction has converted a 10% initial measurement error into a possible error of 1,990% in the answer. This means that the answer would be useless for all practical purposes.

The range of measurement error involved in this realistic example is far greater than the answer itself. The measurement error relative to the original measures has grown out of all proportion due to the ensuing simple calculation. This means that, even in this very simple analysis, the answer is largely composed of error, and should not be trusted. Most datasets and analyses used in social science research are far worse than this. But we hardly ever know because we cannot track the errors from their source.

This propagation of initial measurement errors is usually ignored in training texts, because it cannot be addressed by traditional statistical analysis (i.e. through the archaic use of significance tests). It has become a hidden issue.

As a general rule, fewer and simpler calculations make the propagation of initial errors easier to track, and less likely to occur. Compare, for example, the post-intervention-only and the pre- and post-intervention experimental designs (Chapter 9). In the post-intervention-only design, two similar estimated scores (the means for each group) are subtracted to create an answer that is considerably smaller yet contains the error component of both initial scores.

This means that the maximum relative error will propagate, sometimes very considerably as in the example above, even where the initial errors are within reasonable bounds. But the situation is even worse for the pre- and post-intervention design. First the pre- and post-intervention scores are subtracted for each individual. This creates an answer that is considerably smaller yet contains the error component of both initial scores. These error-ridden gain scores are then averaged, and then the whole process happens again. The two very similar mean gain scores for each group are subtracted to create a difference that is now very much smaller than any of the initial scores, but the maximum error components of each are still added into the computed result. We have no reason to assume that averaging the individual gain scores cancels out all or even any of the initial error. This has been confirmed both by algebraic proof and by repeated simulation (Gorard, 2013a).

So, although error propagation is a problem for any real-life situation where computations are conducted with measures that are less than 100% accurate, it is less of a problem for the simplest analyses where less computation is needed. Do not use complex analyses, and do not dredge your data for apparently exciting results – they will probably be spurious.

## Exercise 14.1

If we are comparing two summary measurements (the average waiting time in two doctors' surgeries perhaps) then each must be assumed to have an error component. Some patients' waiting times may be missing, some may be misestimated, and some may be misrecorded. Imagine that each average is only 90% accurate. Whatever their estimated value is, the actual average waiting times could be up to 10% different. Technically this means that the maximum relative error in each measurement (the error relative to the measurement) is 10%. If the first measurement is 20 minutes and the second is 25 minutes, what is the *maximum* relative error in the difference between them (i.e. the error relative to the achieved result of 5 minutes' waiting time)?

## TAKING REPEATED MEASUREMENTS FOR RELIABILITY

The final section of this chapter considers the notion of the **reliability** of numeric measures (as opposed to the kind of **validity** of measurements discussed so far). 'Reliability' is a rather ambiguous term. It refers to the extent to which once a measurement is taken it can be repeated successfully. This is a useful idea. If we measure someone's height, and immediately measure it again, then this begins to give us a sense of how reliable (repeatable) this measuring process is. If we measure the height in one way (or using one measuring device) and then repeatedly with other devices, then we can begin to assess the reliability of the devices. Or, if we get different people to take the same (ostensible) measurement then we have evidence of the reliability of different observers of the same thing. These approaches all make sense, even though they lead to different measures of reliability. When authors discuss reliability it is important that they specify which of these three types it is.

**Image 14.1** Measuring measures

Whichever type of repeated measurement is used, its reliability can be expressed in different ways. You could look at the range of measurements in practice. The bigger the range of values for one attempted measurement, the lower its reliability and therefore its usefulness. Even if we have a 'valid' way of measuring something, this is irrelevant if we cannot measure it reliably. Reliability does not always imply accuracy. If we measured heights with a bad ruler that was 2 cm too short, then every time we used that ruler it could give consistent/reliable but wrong results. However, an accurate measure does need to be reliable. Reliability could be the level of agreement between tries, and can be expressed quite succinctly, as in the examples in Chapter 20.

Sadly, though, in social science, 'reliability' is most often used to refer to something very different. You will meet it in descriptions of questionnaires and the construction of composite indicators, or indices, often used to try and measure attitudes (or similar). In these questionnaires the respondent is faced with a battery of questions, all of which are used to assess what is basically the same underlying (or latent) variable. In essence, the same or very similar questions are asked repeatedly, and the answers are used in combination to create an overall attitude score. 'Reliability' here is used to refer to the extent that the respondents give compatible responses to all of the questions that are going to be aggregated.

One problem with these constructs such as attitudes and preferences is that there is no explicit thing – the correct answer – with which the purported measure can be compared. Therefore, it is often hard to see attitude scales as real measures of anything. In the absence of the ability to calibrate questions by comparing responses with reality, users and developers of attitude questionnaires have seemingly become obsessed with a version of reliability across different questions. In some ways this is just the same as the reliability of different processes or measurements. We could, for example, have a questionnaire that asked respondents for their date of birth and then for their exact age. Comparing responses to these two items would give a measure of how reliable respondents' answers were. Where the responses match (are consistent) they are reliable. But where they differ there is no way of telling from the two items themselves which if any is the correct one. It is important to recall this in what follows.

So asking what is effectively the same question more than once does not lead to greater accuracy of results. If we asked only one question (e.g. date of birth) we must use the only response. It may be the correct date or it may not. Asking the same question in another way (e.g. exact age) now gives us two responses. In so far as they yield a consistent answer, we have gained no new knowledge of age, and have wasted the time of the respondent, and probably annoyed them in the process. We may even have reduced the response rate (see Chapter 12). On the other hand, it is hard to see what to do with the responses if they are different. We cannot use the average of the two ages. That is very unlikely to be correct, and is not the age/date the respondent gave us for either question. Usually in attitude scales such items are averaged and given equal weight for no good reason (Nunnally, 1975). There is no reason to believe that the true age would fall exactly halfway between two erroneous responses. We could ignore one of them, but then we may as well have only asked the question once. Asking exactly the same question gains nothing, but it does damage our instrument.

Despite this, and despite the existence of good and tested single-item questions for most attitudes (Gorard et al., 2017), attitude questionnaires generally use not one or two items but a whole battery of questions to ask repeatedly about the same thing. Everything said so far about asking a question twice is also true of asking a question three or more times. The key point is that if the items are different, and are actually measuring different things, or different aspects of one thing, then they cannot be used to assess reliability. If they are measuring exactly the same thing again and again, then they are both pointless and dangerous.

One claim in response to this is that asking similar questions repeatedly leads to greater accuracy since random errors in responses are eliminated. This would clearly not be true, even if we assumed that any errors in response were random. We must also assume that a majority of respondents are able to answer any question correctly. If, on the other hand, a majority of respondents are not able to answer a question correctly then the survey is doomed from the start, and the use or non-use of constructs is irrelevant.

Imagine, though, that the errors are random, and we asked 100 respondents about their age (as above), and a random 10 respondents made a random error in their answer. We have 90 correct ages, and 10 wrong ones, although we would not know which is which.

Now we ask the same question again in a slightly different way (as a date of birth), and again a random 10 respondents made a random error in their answer. On average, 9 of the 90 who gave us a correct answer first time will give an incorrect answer the second time (it is random), and 1 person out of 10 will give us a second incorrect answer. So we now have 81 consistent responses, and 19 sets of responses that have at least one error. And we will not know which of any two inconsistent responses is the right one. We are worse off asking the question twice than if we asked it once and got 90 good answers (Gorard, 2010a).

## CONCLUSION

This chapter highlights the fact that most numeric datasets will contain non-random errors (of measurement or recording, missing data and so on), and that these must be assessed and dealt with accordingly. Errors in measurements will matter when you come to do your analysis. The key message is that you must assume your data has errors in it, and remember that this error component will tend to get worse as you conduct your analyses. So do the simplest analyses possible, and do not just assume that small differences or slight trends are noteworthy (see Chapter 8). Other key messages are that you must specify what kind of reliability you mean when you refer to the reliability of your findings, and that merely asking what is meant to be the same question repeatedly will tend to damage your findings.

### Notes on selected exercises

#### Exercise 14.1

The first surface measurement is 20, but the real value could be between 18 and 22 (i.e. out by 10% of 20, or 2, in either direction). The second surface measurement is 25, but the real value could be between 17.5 and 27.5 (out by 10% of 25, or 2.5, in either direction). In reality, the maximum difference between the two measures could be 27.5 ÷ 18, or 9.5. So, the maximum error relative to the achieved answer of 5 is 9.5/5, or 190%. A realistic and acceptable error of 10% in the initial data yields an error of 190% in the research result. This error propagation is an unacknowledged problem in research using numbers.

## Suggestions for further reading

An informative book on measurements and their uses:
Bradley, W. and Shaefer, K. (1998) *Limitations of Measurement in the Social Sciences*. Thousand Oaks, CA: Sage.

More on errors in measurements, when analysing experimental results:
Gorard, S. (2013a) The propagation of errors in experimental data analysis: A comparison of pre- and post-intervention designs, *International Journal of Research and Method in Education*, 36(4), 372-385.

More on the nature of measurements, if you want to examine this idea further:
Berka, K. (1983) *Measurement: Its Concepts, Theories and Problems*. New York: Springer.

# Part IV
# Modelling with data

# 15
## CORRELATING TWO REAL NUMBERS

### SUMMARY

This part of the book moves on to modelling and multivariate analyses. It starts with the idea of correlation, which underlies much of the rest of the book. This chapter describes how to conduct a correlation analysis with two or more variables, illustrating the technique with both real-life and imaginary examples, and showing its benefits and limitations. This approach might be used to address research questions like:

- How strong is the link between earned income and giving to charity?
- Is earning more at age 30 linked to longer life expectancy at age 60?
- Do people who score well in a test of language also tend to score well in a maths test?

### INTRODUCTION TO CORRELATIONS

In Chapter 4 we looked at the cross-tabulation of two categorical variables. In Chapter 5 we looked at the differences in the mean scores of one real number separated into groups in terms of one categorical variable. Each type of variable was explained in Chapter 2. This chapter completes the set of simple bivariate analyses, by considering a comparison between two real number variables, known as a correlation analysis. It also introduces ideas for use in the multivariate analyses to follow.

As in so many examples, the best place to start when considering two real numbers is to plot a graph. Here the graph would be of the relationship between two real number variables, each plotted on one of the two axes. Imagine a dataset of 100 complete cases that record individuals' height and the length of their index finger. We would expect there to be some kind of relationship, such that taller people tend to have longer fingers, and vice versa. A scatterplot would look like Figure 15.1.

---

To draw a scatterplot graph, go to "Graphs" on the SPSS menu, and select "Chart Builder". In the bottom right of the window select Scatter/Dot and drag the first example on the right to the Chart preview. Then from the variables list at the top right of the window, select the variable Height and drag it to the x-axis in the chart preview rectangle. Then drag the variable Length to the **y-axis** in the chart preview. Click OK.

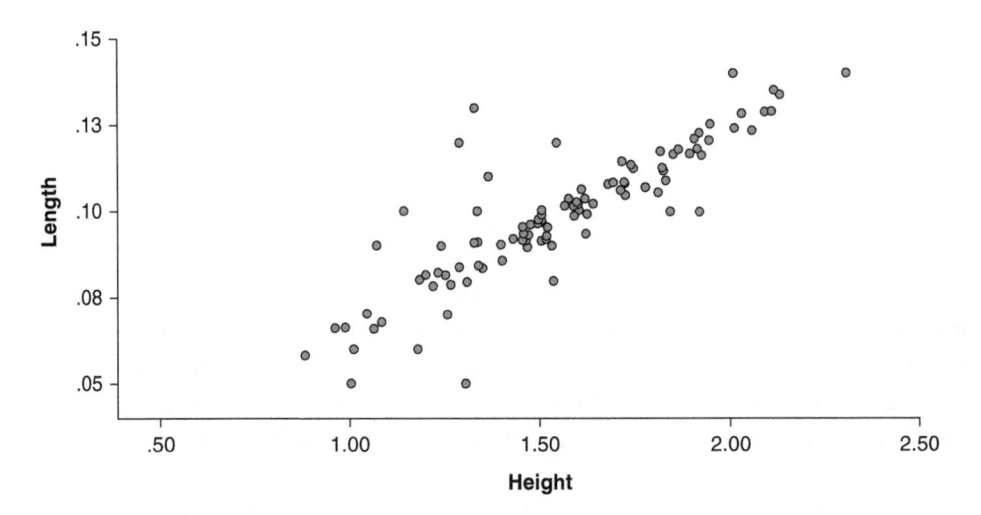

---

**Figure 15.1** Graph comparing height and finger length (imaginary data)

As expected, this shows quite a strong relationship. People with less height (at the left end of the x-axis) tend to have shorter fingers (at the bottom end of the y-axis). People with longer fingers (at the top of the graph) tend to be taller (towards the right of the graph). The relationship is like a straight line but has some discrepancies. A few people have longer fingers than their height would suggest, and some have shorter fingers. We could use this graph to help guess someone's height from their finger length, by finding the length of their finger on the y-axis and reading across until we get to a cluster of dots. Then, reading down, we can find the nearest value for their height. For example, someone with a finger length of 10 cm will probably have a height of around 150 cm according to these made-up figures. This will not be a perfect process, because each length may have more than one height close to it. And the same would apply if we were trying to guess finger length from height.

We can begin to be more accurate than reading off the graph, by conducting a correlation analysis, here creating a Pearson's $R$ correlation coefficient (Pearson is the person who helped devise the analysis, and $R$ is name of the output, or coefficient).

Go to the "Analyze" menu and select "Correlate" and then "Bivariate" (because you are comparing two variables at a time). Move the two variables to the Variables box. Click OK.

```
CORRELATIONS
/VARIABLES=Height Length
/MISSING=PAIRWISE.
```

Analytical software will produce a table of output looking like this, for the variables Height and Length:

| Correlations | | Height | Length |
|---|---|---|---|
| Height | Pearson Correlation | 1 | .866** |
| | Sig. (2-tailed) | | .000 |
| | N | 100 | 100 |
| Length | Pearson Correlation | .866** | 1 |
| | Sig. (2-tailed) | .000 | |
| | N | 100 | 100 |

**. Correlation is significant at the 0.01 level (2-tailed).

As ever, we need to clean this up a bit before thinking about what it means. Any asterisks (here **) relate to significance testing, which we are not interested in (see Chapter 6). The repeated row 'Sig. (2-tailed)' also refers to significance testing and can be ignored. The number of cases (N) is 100 and does not need to be repeated.

So a simpler version of the output would be as in Table 15.1. What the result shows is that each variable is 'correlated' with itself with a correlation coefficient or $R$ of 1 (perfect or 100% correlation obviously). This is a trivial point. More interestingly, our two variables have a correlation with each other of +0.87. This is a clear correlation, because it is quite close to 1. It means that there is a positive linear relationship between height and length. As height increases so does length, as was already clear in Figure 15.1. Without using a table, we could simply report that height and length correlate with a coefficient of +0.87 (to two decimal places).

**Table 15.1** Pearson correlation of height and length

| | Height | Length |
|---|---|---|
| Height | 1 | 0.87 |
| Length | 0.87 | 1 |

$N = 100$

The value of R will always be between −1 and +1. A figure of +1 means that the variation in the two values is identical, and they would **cross-plot** in a graph as a perfect straight line rising in value from left to right (like Figure 15.1 but even neater). An example would be temperatures in Celsius and their equivalents in Fahrenheit, which are just different ways of expressing the same measurement. This would be trivial. There are few examples of perfect correlation between two genuinely different measurements in social science, and an R of +1 probably just means you have measured what is effectively the same thing twice, using different approaches, like people's birth date and their age.

A figure of −1 also means a perfect correlation, but this time it is negative, and the cross-plot would show a neat line decreasing in value from left to right (a mirror image of Figure 15.1). As one value increases, the other decreases in perfect alignment. A correlation of −1 is as strong as a correlation of +1. Both relationships would show a perfect straight line on a scatterplot, but they would be sloping in different directions.

A value of 0 for R would mean that there is no relationship between the two variables, and the cross-plot would simply show a random scatter like that in Figure 15.2.

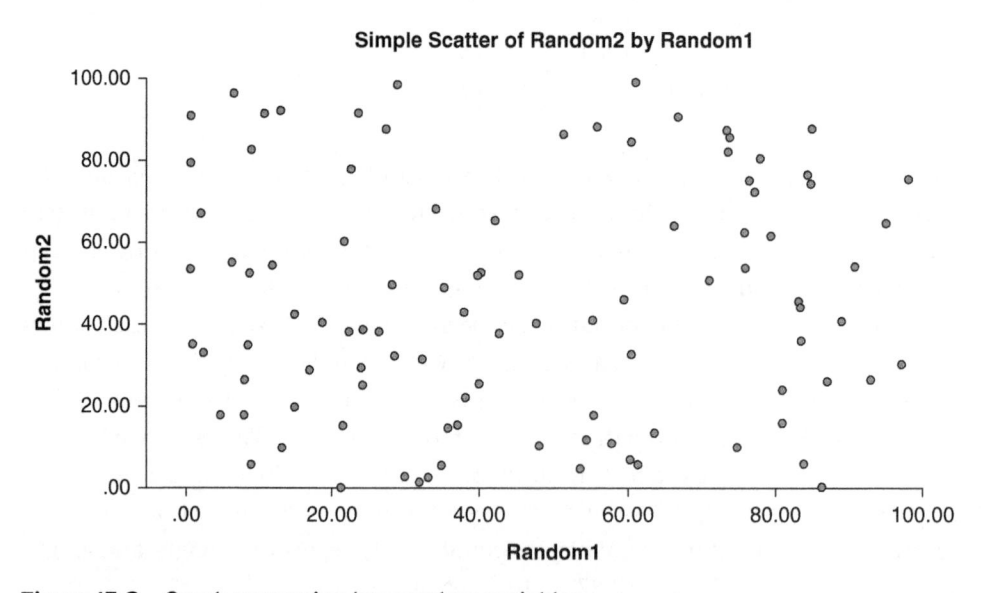

**Figure 15.2** Graph comparing two random variables

Figure 15.2 can be created using the same steps as for Figure 15.1, but with two different variables (here called Random1 and Random2). The 100 uniform random numbers were created as described in Chapter 11. Similarly, the correlation analysis for these two sets of random numbers can be performed as above, with the two new variable names. This could produce the following output:

| Correlations | | | |
|---|---|---|---|
| | | Random1 | Random2 |
| Random1 | Pearson Correlation | 1 | .045 |
| | Sig. (2-tailed) | | .660 |
| | N | 100 | 100 |
| Random2 | Pearson Correlation | .045 | 1 |
| | Sig. (2-tailed) | .660 | |
| | N | 100 | 100 |

Ignoring the clutter, this shows that $R = 0.045$. This is not exactly zero, and $R$ is unlikely to be exactly zero for any two real sets of numbers, just by chance. But it is probably small enough for us to ignore here (especially as we know it is a random result!). Both the graph and the correlation analysis suggest no linear relationship between the variables Random1 and Random2.

To see quite how small 0.045 is, square it. The result is 0.002. This would be the value of $R$ squared, or $R^2$, which is a kind of effect size. See Chapter 5 for more on effect sizes, and Chapter 17 for more on $R^2$. Here it will be sufficient to point out that $R^2$ has a useful meaning – it is the proportion of the variation (strictly the variance, or square of the standard deviation) that is common to both variables. It ranges from 0, again meaning no relationship, to 1, where the two variables are really identical (and again like equivalent temperatures in Celsius and Fahrenheit). It cannot be negative because, even if $R$ is negative, the act of squaring it makes it positive. If two variables have a correlation of +0.5 or −0.5, then 25% of their variance is common between them (and, to that extent, may be measuring the same thing).

### Exercise 15.1

Two professors have done some research and found a strong correlation (+0.69) between people's ability to perform a skilled task and their confidence in conducting such a task. Participants who are more competent are also more confident and vice versa. However, the professors are arguing about what this result means. One professor says confidence is fundamental to carrying out a skilled task, and so it is this confidence that produces the better skills. The other professor disagrees completely and says that confidence cannot create skill. It is clear, to this professor, that the confidence comes from the ability to complete the skilled task well. Which one is the correct interpretation of the correlation?

## A REAL EXAMPLE OF A NEGATIVE CORRELATION

Here is an example of a negative correlation, based on real data (Figure 15.3). The original dataset is the National Pupil Database for all students and schools in England, containing information on student backgrounds and their attainment (here their total points score at Key Stage 4, the national assessment at age 16). The student background data has been aggregated to create a new variable – the number of years while at school in which they were officially recorded as living in poverty. The analyses below are based on the school-level results – the average KS4 scores (examinations at age 16) in each school, and the average number of years students have lived in poverty (eligible for free school meals) while at each school. For convenience, the results are only based on schools in the economic region of the North-East (to permit readers to see the dots on the cross-plot more clearly).

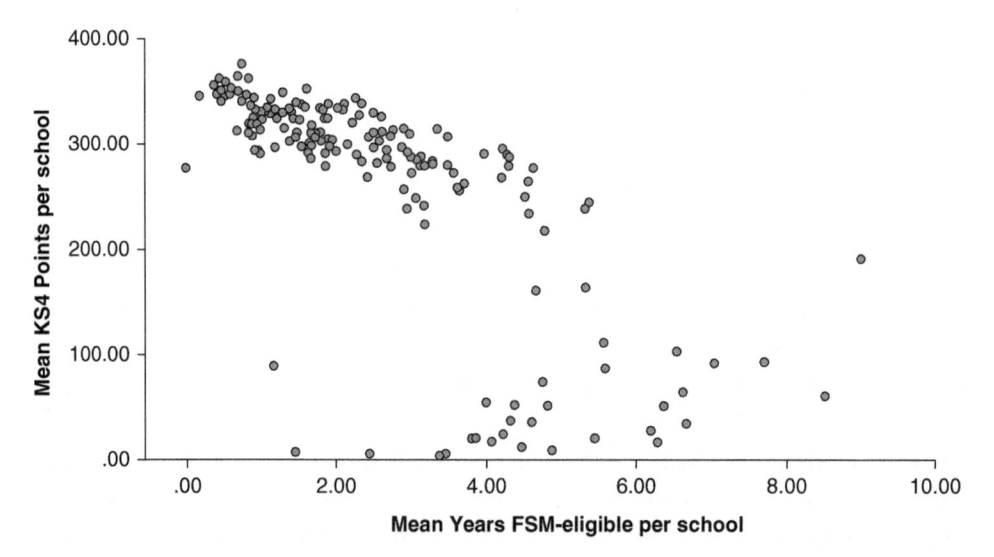

**Figure 15.3** Graph comparing mean years at school living in poverty and mean school attainment, Key Stage 4, North-East England

Figure 15.3 shows a very approximate linear relationship. This lack of complete pattern is usual in real research. There are a few schools whose students have near-zero average attainment scores. These are likely to be small special or hospital schools. Ignoring these, there is a fairly clear negative relationship between the averages for living in poverty and attainment. Schools with the shortest average duration of student poverty have the highest average attainment, and the schools with the longest duration of student poverty (at or near 10 years of poverty while at school) have very low average attainment. This suggests that a Pearson's *R* correlation analysis would be appropriate.

```
CORRELATIONS
/VARIABLES=KS4FSMyearsmean KS4GCSEpointsmean
/PRINT=TWOTAIL NOSIG
/MISSING=PAIRWISE.
```

This produces the following output. Ignoring the clutter, it shows us that 26,816 students were involved in analysis (the data is held for individuals, so each dot on the graph represents the average for all students in each school). More importantly, it shows us that R is –0.705. This is a reasonably strong negative correlation, as expected from the graph, meaning that as one variable increases, the other tends to decrease. Schools with more long-term disadvantaged students tend to have weaker attainment results. Converting the R to $R^2$, this means that 50% of the variation in school attainment is in common with the variation in school mean duration of student poverty. When you first come across this kind of figure in social science it is quite shocking, and makes you want to know why it occurs. It is well known that poverty and attainment at school are negatively related, but what this example illustrates is the growing importance of considering long-term poverty as something rather different from short-term poverty (Gorard and Siddiqui, 2019).

| Correlations | | | |
|---|---|---|---|
| | | KS4FSMyearsmean | KS4GCSEpointsmean |
| KS4FSMyearsmean | Pearson Correlation | 1 | -.705** |
| | Sig. (2-tailed) | | .000 |
| | N | 26816 | 26816 |
| KS4GCSEpointsmean | Pearson Correlation | -.705** | 1 |
| | Sig. (2-tailed) | .000 | |
| | N | 26816 | 26816 |
| **. Correlation is significant at the 0.01 level (2-tailed). | | | |

## MULTIPLE CORRELATIONS

It is possible to correlate as many variables as you like at the same time. In software, instead of simply moving two variables to the Variables box, try moving three or more for one analysis. For example, simply add the variable Age to the example with 100 cases at the start of the chapter.

```
CORRELATIONS
/VARIABLES=Height Length Age
/MISSING=PAIRWISE.
```

The cleaned-up output is shown in Table 15.2. As before, Height and Length correlate with each other, with an R of +0.87. Height also correlates with Age with an R of +0.97, and Length correlates with Age with an R of +0.86. As expected, these are all large correlations. In their early years (perhaps up to age 21), older people tend to be taller than younger ones, and to have longer fingers.

**Table 15.2**  Pearson correlations of height, length and age

|        | Height | Length | Age   |
|--------|--------|--------|-------|
| Height | 1      | 0.866  | 0.974 |
| Length | 0.866  | 1      | 0.860 |
| Age    | 0.974  | 0.860  | 1     |

N = 100

Note how easy it is to conduct a multiple correlation. Of course it gets a bit more complicated with more variables, but not by much. And however many variables are involved, the procedure remains the same as here.

### Exercise 15.2

A friend of yours has conducted a correlation analysis with three real number variables, and obtained very, very low values for R (like –0.04 or +0.007). So in their report they say that this result demonstrates that the three variables are not related to each other - each $R^2$ is too small to be worth considering any further. Is this the correct interpretation?

## ASSUMPTIONS AND LIMITATIONS OF PEARSON'S R CORRELATION ANALYSES

In order to conduct a correlation analysis to compute Pearson's R you will need two (or more) real number variables, as described in Chapter 2. And you will need to draw one or more graphs in order to check that each pair of variables is at least approximately linearly related (as described above). How close to a straight line this has to be is a matter of judgement for you. If the variables do not form something like a straight line on a graph then the analysis might give a misleading answer of no relationship when in fact the relationship was just not a linear one. For example, the variables might create a curvilinear pattern, or a U-shaped or S-shaped pattern.

If your graph shows a nonlinear relationship such as a curve, you may be able to play around with the scales of one or both variables in order to convert the pattern to a straight line. For example, converting one variable into its logarithm before plotting can convert a curve into a straight-line relationship (e.g. Gorard 2019b). However, using such transformations to overcome nonlinearity might introduce further problems (Harwell and Gatti, 2001), and this issue is not covered any further here. For the moment, stick to clearly linear patterns.

Sometimes, when you are dealing with a very large dataset, there might be so many cases that a cross-plot looks like a dark shadow, as cases plot on top of each other, making it hard to see the relationship. You could randomly select a subset of the cases from your big dataset, and plot these in the hope that the pattern will be clearer. Or you could rely instead on your theoretical understanding in order to decide that there might be a linear pattern. For example, at time of writing I do not know the school examination results for 16-year-olds in England in 2025. But I would be very surprised if there were not an approximate linear relationship between the exam scores of these future students when they were aged 11 and their exam scores when aged 16.

**Image 15.1** Correlation or causation?

A standard caution given about the interpretation of correlation is that it gives no indication of the real nature of the relationship between the two variables. There may be a clear correlation between the number of children in any house in Denmark and the number of storks nesting on its roof. This is not evidence, necessarily, that the activities of storks are involved in childbirth. It is perhaps more likely that larger houses tend to have both more roof space and more residents. If children with bigger feet tend to spell better it is not likely that bigger feet are involved in spelling, or that being able to spell makes your feet grow. More likely, older children tend to spell better and also have bigger feet, on average. In the following chapters, and whenever you read about correlation or regression and causal claims, bear this key point in mind. Correlation does not necessarily imply the existence of a causal relationship.

Some resources and writers provide guidelines about what counts as a 'strong', 'medium' or 'weak' correlation. But these are fairly arbitrary and not very useful in practice. Instead, use your own judgements based on the dangers of being wrong, and so on (as explained in Chapter 8).

## A REAL EXAMPLE OF A SMALL CORRELATION: THE SUMMER-BORN PROBLEM

Here is a real example in which the correlation is so small that many texts and commentators would say that it should be ignored. In England, children generally attend school with an age cohort of whom the oldest was born on 1 September of one year, and the youngest was born almost a year later, on 31 August the following year. The precise age of a child within their school year has been shown to be strongly linked to attainment, later life outcomes, and wider personal development (Gorard, 2018). In England, only 49% of the younger, summer-born children who start school in September achieve a 'good level of development' in their first year, compared with 71% of autumn-born pupils (Department for Education, 2016). Figure 15.4 plots the average point scores that all pupils at school in England in one year achieved in their Key Stage 2 examinations at age 11, against the month of their birth. It shows that younger pupils have lower attainment, on average, and that their attainment declines with every month that they are younger than the oldest in their year.

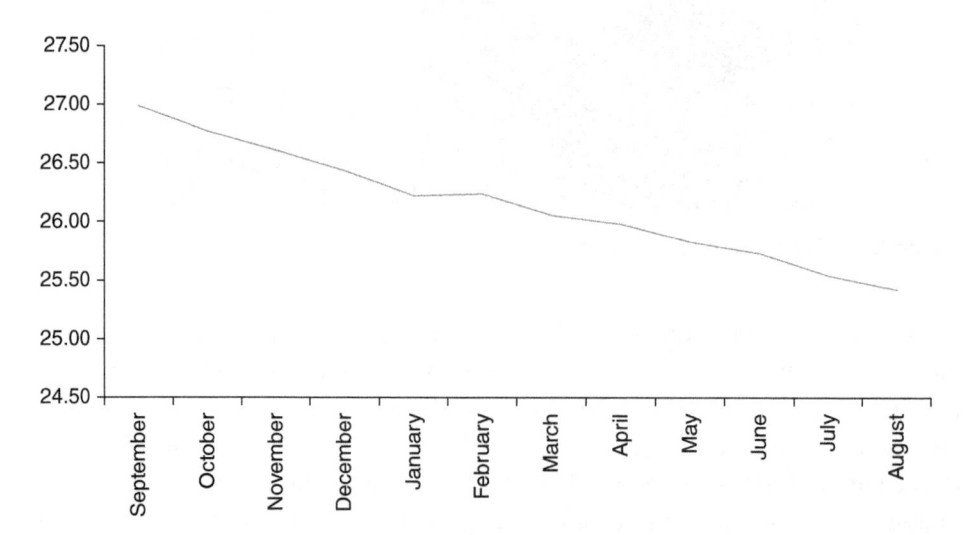

**Figure 15.4** Attainment in terms of Key Stage 2 points, by age-in-year, England

Note that because month of birth is a category, this graph should really be a scatterplot (as above). However, with so few dots it is easier for a reader to see as a line. Similarly, the origin of the graph has been made not zero, in order to help readers see the otherwise very gentle slope. Both are OK on occasion to help the reader, as long as the reader is aware of what is being portrayed.

This graph portrays a small negative correlation between age-in-year and attainment at school. The Pearson's $R$ coefficient is –0.04. This is small. The value of $R$ is way below the level that traditional texts suggest has any meaning. But such texts could be wrong. The younger children in any cohort at school are less likely to pass the entrance test for a grammar school, or be entered for any public examinations, and are 10% less likely to go to university than their older peers. By the age of 18, they have had 12 or more years as the youngest, least mature, and maybe the smallest, person in their year. The summer-born pupils are less likely to be picked for competitive sports, more likely to be bullied, tend to be less happy at school, have lower self-esteem, and are rated as lower ability both by teachers and in tests (Crawford et al., 2011). This is an international phenomenon (Ballatore et al., 2016; Melkonian and Areepattamannil, 2017), and it continues up to and beyond university age (Abel et al., 2008).

Younger children are also more likely to be labelled as having a special educational need (SEN) or learning challenge. Figure 15.5 shows that the labelling of students as SEN, supposedly by an expert, is clearly related to their age-in-year. SEN labelling is over 60% more common for the youngest pupils than for the oldest in any school year.

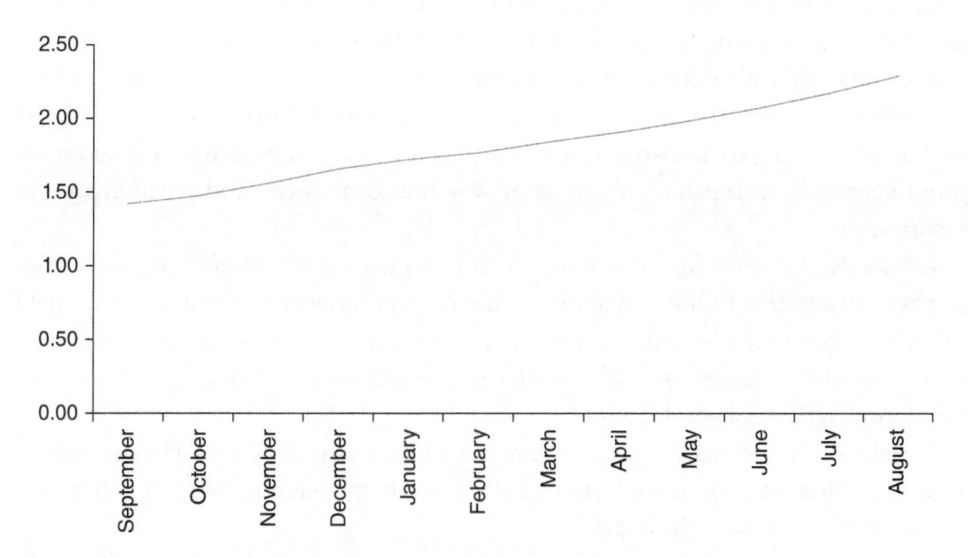

**Figure 15.5** SEN reporting by age-in-year, England, 2015 KS4 cohort

This is absurd. There is no reason to believe that genuine SEN identification should be distributed like this. Therefore, the differences are most likely to be just an artefact of the key dates in the education system. It is probable that children with lower results at school tend to be considered for SEN labelling more often, and that their precise age-in-year is not being taken into account sufficiently by the experts making the decision. Younger children in any cohort will tend to struggle more on average, and so become visible as apparent 'underachievers'. SEN then provides a mistaken reason for this. Even more seriously, these younger children are leaving school with worse qualifications, for no apparent reason other than being younger. A small correlation (effect size) *can* have substantial consequences in real life.

## Exercise 15.3

There is a well-known saying that correlation is not the same as causation. So does this mean that a correlation tells us nothing about a possible causal link between two variables? If not, why not?

## CONCLUSION

A simple correlation, used properly, can tell you a lot. A correlation is the start, not the end, of an investigation, and its explanation is likely to involve theoretical considerations, and the triangulation of knowledge from other data sources. Even then the proposed explanation can only be a tentative one. This basic fact is important to note before we consider more complex designs. There is a danger that novices confuse complexity with rigour, whereas the more complex designs mentioned in later chapters suffer from the same flaws and limitations as the simplest correlation, as well as introducing their own additional problems. In fact complexity and rigour are often negatively 'correlated', in the same way that complexity and reliability are in engineering.

The kinds of correlations described in this chapter are based on deviations from the best-fitting straight line. Therefore, using the techniques described here with data exhibiting curvilinear or other relationships is unlikely to be effective. Alternative correlational techniques are available for curved relationships (Norušis, 2000) and for categorical variables (Siegel, 1956).

Correlation is the basis of many more advanced techniques for analysis, such as factor analysis, regression and structural equation modelling, some of which are described in the following chapters.

## Notes on selected exercises

### Exercise 15.1

Neither professor is correct, in the sense that a correlation does not imply any kind of direction. Correlating variable A with B gets the same result as correlating variable B with A. Also a correlation is just a pattern. It does not, in itself, mean A and B are causally linked to each other (in either direction). Either professor could be right, or each of the two variables might assist the other, or both could be produced by some other factor (such as age or practice).

## Exercise 15.2

No. This interpretation is not justified by the available evidence (yet). First, and most obviously, the correlation analysis can only examine near-linear relationships. Your friend might be justified to say that there is no evidence of linear relationships, but there could be evidence of some other kind of perfectly valid relationship (like a U distribution). This is one of the reasons why we should start the analysis with a series of bivariate cross-plot graphs to get a sense of how things look. Second, your friend needs to use judgement. Although this is unlikely, even a small effect size could be important if the potential gain in knowledge from the study was very great (as in the summer-born example). There is no standard scale for the substantive importance of correlation coefficients. See Chapter 8.

## Exercise 15.3

There are two ways in which the standard caution about interpreting a correlation analysis as having no relevance to a cause-effect model is overstated. A correlation does not imply that the two variables are necessarily cause and effect (or in which direction). But a cause-effect relationship must show a correlation when it is analysed. Correlation is one element used to help identify a causal model (Gorard, 2013b). This means that a correlation analysis can show that a causal model does *not* exist. If two variables do not correlate then neither can be the cause of the other. People often forget the possible importance of such a negative result. If paying employees more were not correlated with better performance at work, for example, this could be an important, if perhaps unpopular, finding!

## Further exercises to try

Find a suitable real-life dataset you are interested in (or use one provided on the website). Select two real numbers, and run a correlation analysis. Does the finding (*R*) suggest that there is a strong linear relationship between them? Is the relationship negative or positive? Cross-plot the two variables to see if this helps to explain your result.

## Suggestions for further reading

A chapter introducing ideas about methods for correlation:
Gorard, S. (2021) Simple statistical and correlational research. In R. Coe, L. Hedges, M. Waring and L. Day-Ashley (eds), *Research Methodologies and Methods in Education*, 3rd edition. London: Sage.

A book challenging the idea that correlation is not causation:
Pearl, J. and Mackenzie, D. (2018) *The Book of Why: The New Science of Cause and Effect*. New York: Basic Books.

# 16
# PREDICTING MEASUREMENTS USING SIMPLE LINEAR REGRESSION

## SUMMARY

Chapter 15 introduced the idea of a correlation between two variables, in which each variable tends to change values in step with the other, but without being either identical, or the antithesis of the other. There is a positive correlation between people's height and the length of their index finger, so that taller people will tend to have longer fingers. But the match is far from perfect.

It is also possible to predict the value of one variable from the corresponding value of the other variable in a correlation, to some extent. One way of doing this is known as regression, and regression modelling is the topic for this chapter. This is our entry point into the world of the statistical modelling of social events. The focus here is on modelling with the data that we have, rather than with theoretical data we might have had, or that could exist. This is the clearest and safest approach (Berk, 2010).

The approaches in this chapter could be used to address research questions such as:

- Is length of prison sentence a good predictor of the likelihood of reoffending?
- How good is the number of years of education as a predictor for income?
- How useful is precipitation (or chance of rain) as an indicator of daily temperature?

## SIMPLE LINEAR REGRESSION

We will start with consideration of simple linear regression. In order to carry out the simplest form of regression analysis, when predicting one measurement from another,

# PREDICTING MEASUREMENTS USING SIMPLE LINEAR REGRESSION | 195

we need two real number variables, just as with correlation. For the first example, we will use the imaginary data on height and index finger length from Chapter 15. Both are clearly real number variables (Chapter 2). Just as with Pearson's $R$ correlation, we would also need some evidence that these two variables are linearly related to some extent. This means that their relationship must look a little bit like a straight line, so that as one variable increases in value the other either increases or decreases in value in a roughly proportionate manner. The easiest way to assess this is to draw a cross-plot of the values of the two variables.

The output will look like Figure 16.1, which simply repeats the cross-plot for our example of height and index finger length (Figure 15.1 from Chapter 15). There is clearly a kind of straight-line relationship between the two variables. The two values correlate at $R = +0.87$ (Chapter 15). If we know the height of a person we could make a good estimate of the length of their index finger just by reading this graph, and vice versa.

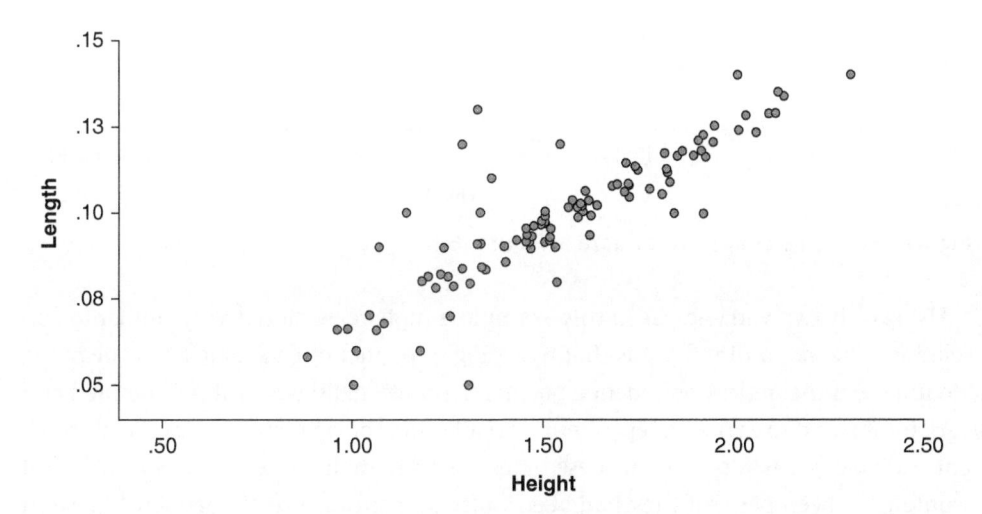

**Figure 16.1** The relationship between height and index finger length (imaginary data)

All that linear regression does is to formalise this pictorial relationship, and the predictions that are possible from it. It does so by creating a line of best fit, which is the line on the graph that minimises the deviations of all points from that line. Or more accurately, this is the line that minimises the squares of the distance between it and all data points (see Chapter 5 for more on the use of squares for this purpose).

---

Draw a scatterplot by following the instructions in Chapter 15 (see p. 182). A line of best fit can be added to a graph by double-clicking the graph in the SPSS output box. This will open it in a Chart Editor window. Go to the "Elements" menu, and select "Fit Line at Total". Click OK.

---

The resulting output above now looks like Figure 16.2. Once this line of best fit has been calculated, it is possible to use it to read off the values of one variable from the values of the other. This reading or prediction of one value from the other will not be perfect, and its accuracy will depend on how close to linear the observed relationship is, or rather how far the observed points deviate from the best line. Any such prediction will contain an error component. What regression does is to try and minimise that error component, so making the prediction as accurate as possible, based on all of the data available.

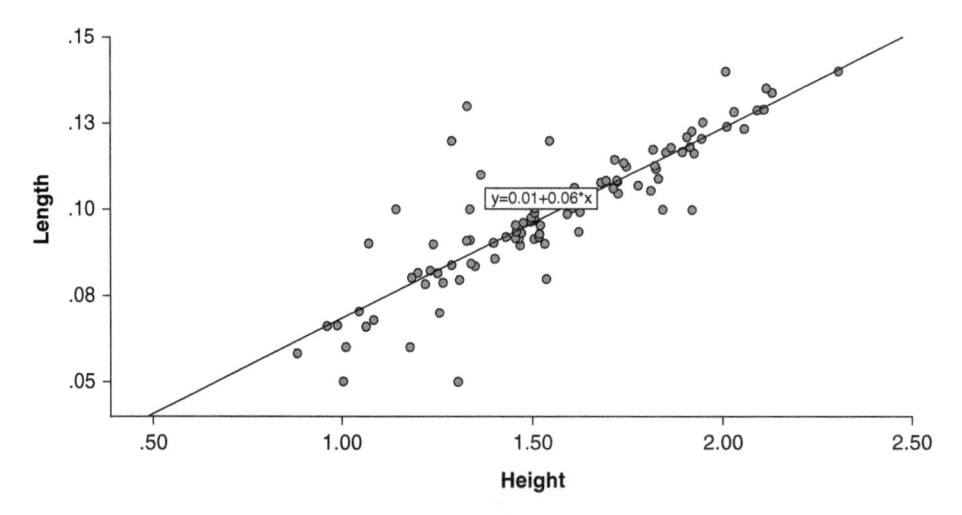

**Figure 16.2** Line of best fit for data in Figure 16.1

Using only two variables, as in this example, simple regression is very similar to correlation. The major difference is that by trying to predict one variable from the other, we impose a theoretical precedence on them. Traditionally you will see the predictor variable referred to as the 'independent' variable, and the predicted one as the 'dependent' variable. We will stick with these names because their use is so widespread, but it would have been better if they had been called something else. 'Dependent' makes it sound as though the 'independent' variable was causally involved in creating the value of the dependent variable. In fact the whole thing is just a prediction of one from the other, and there is no indication, as yet, of a causal connection. We are not saying that the correlation/regression between height and finger length means that height is causing finger length or vice versa. That would be silly. It is crucial to recall this point when regression modelling becomes more complex. Regression is a way of modelling or envisaging your data. It is not in itself a test or demonstration of causal influence.

──────── **Exercise 16.1** ────────

If we try to predict the length of someone's index finger from their height, which is the dependent and which the independent variable?

# PREDICTING MEASUREMENTS USING SIMPLE LINEAR REGRESSION

If we had a dataset of height and finger length for 100 people, with a linear relationship as in Figure 16.2, we could now run a simple regression.

---

The "Analyze" menu offers a choice entitled "Regression", which in turn offers a choice entitled "Linear". Select this, move the variable Height to the Independent(s) box, and the variable Length to the Dependent box. Click OK. Or:

REGRESSION

/MISSING LISTWISE

/STATISTICS COEFF OUTS R ANOVA

/CRITERIA=PIN(.05) POUT(.10)

/NOORIGIN

/DEPENDENT Length

/METHOD=ENTER Height

---

The output might look like this:

Regression

| Variables Entered/Removed[a] | | | |
|---|---|---|---|
| Model | Variables Entered | Variables Removed | Method |
| 1 | Height[b] | . | Enter |

a. Dependent Variable: Length
b. All requested variables entered.

Model Summary

| Model | R | R Square | Adjusted R Square | Std. Error of the Estimate |
|---|---|---|---|---|
| 1 | .866[a] | .750 | .748 | .00981 |

a. Predictors: (Constant), Height

ANOVA[a]

| Model | | Sum of Squares | df | Mean Square | F | Sig. |
|---|---|---|---|---|---|---|
| 1 | Regression | .028 | 1 | .028 | 294.332 | .000[b] |
| | Residual | .009 | 98 | .000 | | |
| | Total | .038 | 99 | | | |

a. Dependent Variable: Length
b. Predictors: (Constant), Height

| Coefficients[a] | | | | | | |
|---|---|---|---|---|---|---|
| | | Unstandardized Coefficients | | Standardized Coefficients | | |
| Model | | B | Std. Error | Beta | t | Sig. |
| 1 | (Constant) | .013 | .005 | | 2.558 | .012 |
| | Height | .055 | .003 | .866 | 17.156 | .000 |
| a. Dependent Variable: Length | | | | | | |

As ever, this can look scary until you realise that most of it is not relevant here (see Chapter 6). We can ignore the first box after the title 'Regression' for now, and ignore the third box headed 'ANOVA' always. It concerns a significance test which we are not using here. We can also ignore the final column of the second box, labelled 'Std. Error of the Estimate', and the last two columns of the fourth box, labelled 't' and 'Sig.'. We can ignore the sub-column labelled 'Std. Error' in the fourth box. Again, these elements are all part of significance testing which we are not doing here. It would be good to have a switch in software to turn all of this stuff off when not needed (i.e. almost all of the time). We can also ignore, for the moment, the first column in each table (the model number), because we only have one model.

Maybe copy the output to Word from whatever analytical software you have used, and delete the bits that do not matter, or that do not make sense. Simplifying the main results leads to Tables 16.1 and 16.2.

**Table 16.1** Model summary predicting finger length from height

| R | R squared | Adjusted R squared |
|---|---|---|
| 0.866 | 0.750 | 0.748 |

**Table 16.2** Coefficients for model predicting finger length from height

| | Unstandardised B coefficient | Standardised beta coefficient |
|---|---|---|
| Constant | 0.013 | – |
| Height | 0.055 | 0.866 |

This is now much simpler. Shorn of irrelevancies, our regression model has given us an $R$ coefficient of 0.87 (which is 0.75 when squared). This $R$ is, of course, exactly the same as the correlation coefficient we already knew (see Chapter 15). And it means the same thing as it did then. There is a strong positive relationship between the two variables (Table 16.1). Similarly, the $R^2$ effect size of 0.75 means the same as in Chapter 15. Around 75% of the variation in each variable is predictable from the variation in the other (as can also be seen in the strong linear relationship in Figure 16.1). This is a high correlation. The value of $R$ would remain the same if we reversed the independent and dependent variables, trying to predict height from finger length instead. The model, in which we pre-specify which variable is the dependent one, can provide no evidence at all that this variable really is the one *dependent* on the other one.

In regression, however, the model is not just specifying a mutual relationship between two variables. We have asked for it to be used to predict the dependent variable (length) from the independent variable (height). The method of prediction is shown in Table 16.2. We have a choice of how the coefficients are represented. In the first version (second column) we see the unstandardised coefficient, which is relevant to the units for each variable being measured. In Table 16.2 the 0.055 relates to a measurement in centimetres. This is the version best used to make the prediction of finger length. The **standardised coefficient** has been converted to a standard scale so that all such coefficients in any regression are on the same scale, and so their values can be compared directly (see Chapter 17).

The best prediction we can make for any individual finger length using the data in Table 16.2 is that it will be 0.013 (the 'Constant'), plus 0.055 (the '$B$ coefficient') times that person's height. This is the same equation that appears in the centre of Figure 16.2. So, for this made-up example, someone who was 200 cm (2 metres) tall might be expected to have an index finger of around 11 cm (0.013 + 200 × 0.055). Someone who was 1.5 metres tall might be expected to have a finger around 8.3 cm long (0.013 + 150 × 0.055). The constant is simply the point at which a graph like Figure 16.2 crosses the vertical $y$-axis. That is really all there is to a regression model.

---

Software like SPSS will do all of these calculations for you, if you want to know the predicted value of the Length variable for each value of the variable Height. In the Linear Regression dialogue box there is a Save button on the right. Click this and then select Unstandardized Values. When you run the regression analysis, a new variable is created, representing the predicted scores for each case in the dataset, based on the model from Table 16.2.

```
REGRESSION
/MISSING LISTWISE
/STATISTICS COEFF OUTS R ANOVA
/CRITERIA=PIN(.05) POUT(.10)
/NOORIGIN
/DEPENDENT Length
/METHOD=ENTER Height
/SAVE PRED
```

---

You now know how to conduct simple linear regression. It is not much harder than doing a correlation analysis, is it?

## Exercise 16.2

Find a dataset you have access to, or are interested in. Select two real number variables, and draw a cross-plot graph of their relationship. Does it look anything like a straight line? Run a correlation analysis. Does the R result match your expectations based on the graph? Run a regression analysis with one variable as the imagined dependent variable and other as the predictor. Note the relationship to the correlation analysis.

## CONCLUSION

This chapter introduced the idea of a linear regression model. This goes beyond a correlation by predicting as accurately as possible the value for one variable (the outcome or dependent variable) from the other variable (the predictor or independent variable). The true importance of this is revealed in the next chapter that looks at regression models with many predictor variables.

## Notes on selected exercises

### Exercise 16.1

If predicting the length of someone's index finger from their height, then height is an independent variable (a predictor), and length of index finger is the dependent variable (being predicted).

### Exercise 16.2

The more clearly the graph shows a straight-line relationship, the higher $R$ will tend to be (either positively or negatively). The $R$-value should be the same as for a simple bivariate correlation analysis (Chapter 15).

## Further exercises to try

Find a suitable real-life dataset you are interested in (or use one provided on the website). Select a real number variable as the outcome, and another real number as the predictor, in a simple regression analysis. Make sure you only use real number variables. Try using the same two variables in a correlation analysis. Check that the value of $R$ is same as in the regression model.

## Suggestions for further reading

An introduction to correlation:
Chapter 10 in Gorard, S. (2003b) *Quantitative Methods in Social Science: The Role of Numbers Made Easy*. London: Continuum.

A 'how to' book on correlation and regression:
Miles, J. and Shevlin, M. (2001) *Applying Regression and Correlation: A Guide for Students and Researchers*. London: Sage.

# 17
# PREDICTING MEASUREMENTS USING MULTIPLE LINEAR REGRESSION

### SUMMARY

Chapter 15 introduced the idea of a correlation between two variables, and Chapter 16 showed how such a correlation can be used to predict the value of one variable using the other. This chapter extends the idea to multiple linear regression, in which two or more variables are used together to predict an outcome.

The approaches in this chapter could be used to address research questions such as:

- Is income a factor in life expectancy, once social class is accounted for?
- Is parental background or student attainment at school a more important indicator for success in life?
- Is poor-quality housing a symptom, or a possible cause, of other indicators of disadvantage?

### INTRODUCTION

We turn now to the only slightly more complex topic of multiple regression. A very similar kind of linear relationship to that for height and finger length in Chapter 16 might be also true for finger length and age, for people aged up to 21. As children and young people grow older their fingers will tend to grow longer. So a simple regression model could be created in the same way as in Chapter 16, using age as the sole predictor variable, to predict finger length. But what is especially useful about regression is that any number of predictor variables can be used to predict the outcome, within the same model. So we could create just one regression model with finger length as the outcome, and both height and age as predictor variables at the same time. This would be a *multiple* linear regression, and it could make our predictions of length more accurate than using either predictor variable alone.

To conduct a multiple regression we conduct the same steps as for simple regression, but with more independent variables.

---

Go to the "Analyze" menu, select "Regression", and from the sub-menu select "Linear". Move Height to the Independent(s) box, and both the variables Length and Age to the Dependent box. Click OK.

```
REGRESSION
/MISSING LISTWISE
/STATISTICS COEFF OUTS R ANOVA
/CRITERIA=PIN(.05) POUT(.10)
/NOORIGIN
/DEPENDENT Length
/METHOD=ENTER Height Age.
```

---

Ignoring the clutter that the analytical software generates, the output would be as in Tables 17.1 and 17.2. Even though our simple regression model already had a high R-value (0.866), using age as well as height to predict length produces a slightly higher value for $R$ (0.869), with a correspondingly higher $R^2$ (0.755). Again, $R^2$ is an effect size, which suggests that around 76% of the variation in the variable Length can be predicted by the combined variation in the variables Height and Age.

**Table 17.1** Model summary predicting finger length from height and age

| R | R squared | Adjusted R squared |
| --- | --- | --- |
| 0.869 | 0.755 | 0.750 |

As explained further in Chapter 18, just adding more variables to a regression model can create an apparently better prediction, even if the added variables are meaningless. The 'adjusted R squared' in Table 17.1 is an attempt to overcome this problem by taking account of the number of predictors in the model. The more predictors there are, the more $R^2$ is adjusted downwards. Here with only a few predictors it makes little or no difference if you work with $R^2$ or adjusted $R^2$ (but it is always worth comparing the two).

Turning now to the predictive model, this would be in the same format as the simpler example in Chapter 16 (Table 16.2), but with more predictor variables (Table 17.2). The predicted finger length for any case would be computed as the Constant (now 0.028) plus 0.036 times the Height, plus 0.002 times the Age. Any number of variables can simply be added to the model as: their coefficient times their value for each individual. And as in Chapter 16, the predicted value for each case can be computed and saved by your software for you to use in further analyses if you want. The coefficient for Height is smaller in Table 17.2 than it was in Table 16.2, because some of the variation is now accounted for by Age (since age and height are also correlated with each other).

**Table 17.2** Coefficients for model predicting finger length from height and age

|  | Unstandardised B coefficient | Standardised beta coefficient |
| --- | --- | --- |
| Constant | 0.028 | – |
| Height | 0.036 | 0.559 |
| Age | 0.002 | 0.315 |

### Exercise 17.1

Identify three real number variables. Draw a cross-plot graph of each pair. Note any near-linear patterns. Run a correlation analysis for each pair. Run a regression analysis with one variable as the imagined dependent variable and one as the predictor. Note the relationship to correlation. Now run a regression with two predictors. Note the differences to the bivariate regression results.

Several things should be clear from the examples so far. Regression is rather easy and intuitive in nature. It is not a definitive test of anything, and it does not prove anything. It is only of use for estimating, or providing a best guess in the absence of any other evidence. It is impossible to demonstrate causal relationships using regression alone. An individual's height does not cause their finger length or vice versa. It does not matter how elaborate a regression model is, the model is merely an expression of association between the two or more variables.

We may imagine in some regression models that the link is a causal one, but the indications of this causality come from our knowledge of the variables used, and of the real world outside the regression model. They are not from the model itself.

For example, imagine that a student's examination performance can be partly predicted by their repeated absence from lectures. An individual with higher recorded absence from lectures might tend to get lower grades in their assignments. We might imagine that absence therefore impairs performance. But that explanation must be based on our wider knowledge of the mechanisms involved, not on the regression pattern alone. Otherwise, it could be that progressive failure in a course leads to non-attendance or avoidance of lectures, or that the two measures could be mutually reinforcing, or there might be any number of other things affecting both measures (such as illness and lack of motivation). At least implicitly, many examples of regression adopt a 'path analysis' approach to testing the feasibility of a causal path previously determined by other evidence or a theory (Maruyama, 1998).

Any model could be probably be improved by considering further variables. It is likely that student examination outcomes would vary by subject studied, sex, ethnicity, prior test scores, parental occupational group, and many other factors, as well as in terms of absence from lectures. A practical problem is that many of these further variables will themselves be interrelated. We cannot therefore simply total the correlations of each variable with the examination outcome scores.

For example, if a prior test score is correlated with student examination outcomes, we cannot simply add this correlation to the correlation with the number of lectures missed. This is so for two main reasons. First, prior scores and absences are likely to have some correlation between themselves, so using both together leads to our using their common variance twice. The real multiple correlation between prior scores and absences on the one hand and examination outcomes on the other is likely to be less than the sum of the two correlations. Multiple linear regression tries to overcome this problem. It takes into account the correlations between multiple independent variables when combining them, in order to predict/explain the variance in the dependent variable.

Second, the real multiple correlation between prior scores and absences on the one hand and examination outcomes on the other could actually be greater than the sum of the two correlations. This would mean that there is an **interaction** effect between prior attainment and absences, whereby the one variable reinforces the impact of the other (Pedhazur, 1982).

## INTERACTIONS

Multiple regression can consider the interaction between two or more predictor variables, in addition to any correlation of the outcome variable with each individual predictor variable. An interaction occurs when the relationship between the dependent variable and an independent variable changes as another independent variable(s) changes. The chances of getting a job if one is poor might be different for different ethnic groups, for example.

To create an interaction in your model, you can simply create a new variable in your dataset by multiplying the two variables you wish to see interact. Here, we will use Age and Height to create a new variable InteractionAgeHeight (a clumsy name, but one typically generated by analytical software; see Chapter 21).

Go to the "Transform" menu, select "Compute Variable", type the name of your new interaction variable in the Target Variable box (here InteractionAgeHeight), and then move the two variables you are interested in to the Numeric Expression box, and put an asterisk between their names. Click OK.

COMPUTE InteractionAgeHeight=Age*Height.

EXECUTE.

To run a regression using this interaction variable, the process is the same as in the last example of multiple regression, but with three variables now instead of two, adding InteractionAgeHeight to the list of predictor variables.

The results could be as in Tables 17.3 and 17.4. Table 17.3 shows that the overall model (R) is no different than that of the model without the interaction (Table 17.1). This suggests that there is little or no interaction effect between age and height when predicting finger length. Age and height are both good predictors of length, but there is no way in which people who are particularly tall or short for their age have noticeably different finger lengths than other people of the same height, for example.

**Table 17.3** Model summary predicting finger length from height and age, with interaction

| R | R squared | Adjusted R squared |
| --- | --- | --- |
| 0.869 | 0.755 | 0.748 |

This is confirmed by the coefficients in Table 17.4. Adding the interaction variable makes no clear difference to the model. The other coefficients remain pretty much as they were in Table 17.2 with the simpler model, and the coefficient for the interaction term is quite near zero.

**Table 17.4** Coefficients for model predicting finger length from height and age, with interaction

|  | Unstandardised B coefficient | Standardised beta coefficient |
| --- | --- | --- |
| Constant | 0.025 | – |
| Height | 0.038 | +0.589 |
| Age | 0.002 | +0.382 |
| InteractionAgeHeight | 0.000 | −0.097 |

Adding the interaction term is not justified here, and so it is preferable to stick to the simpler model that is just as good at predicting the outcome. Remember 'as simple as possible' is our friend when doing research. This kind of null result from an interaction is common. I have rarely found interactions to make much difference compared to the complexity they bring with them. Only worry about interactions when your idea suggests that they are especially appropriate, or where a research question or theory requires that an interaction is tested. Certainly do not interact all variables with all others just to see if any can increase your $R$. This would be dredging, and would increase the chances of being misled by small, fluke patterns (as would be represented by a lower adjusted $R^2$).

## CREATING THE SIMPLEST MODEL: A REAL-LIFE EXAMPLE

As discussed throughout this book, analyses and their reporting should be as simple as possible, while remaining valid. In regression models this means that we want to present the most suitable model for our research question, which explains the most variation in the dependent variable, but uses the fewest possible predictors to do so. This section explains part of how to achieve that ideal.

The model you get if you simply enter all of the variables into regression at once will generally explain the most variation in the dependent variable. However, it is possible to create simpler models containing fewer variables but still explaining the same or nearly the same amount of variation. These models are easier to use and understand, and so are more practical.

For example, it may be that ethnicity and first language, as variables, are measuring much the same thing when acting as predictors of labour market outcomes. The same may be true of social class and indicators of poverty as predictors of attainment at school. In such cases, we may be better off picking the best single indicator from a group of related measures, and using only that one. We could pick the best indicator on theoretical grounds, or in terms of availability. Both of these approaches are fine. However, the most common reason for selection of variables is the proportion of variation that they help to explain. If first language and ethnicity are related, and language is a better predictor of labour market outcomes, then we might omit ethnicity from our analysis.

This idea is illustrated using a real-life dataset from a study evaluating whether a school intervention called Switch-on Reading helped struggling young readers to read better than if they were taught otherwise (Gorard et al., 2015). The same dataset is considered from other perspectives in Chapters 5 and 12. A group of 308 pupils aged 11 were individually randomised either to be given one individual lesson per week with a teaching assistant trained in Switch-on Reading, or to remain in the usual lesson. After one term, the outcome of interest was the reading test scores for all pupils. Possible explanatory variables included whether each pupil received the intervention or not, the number of individual sessions they had, their reading test scores before any of them had the intervention, and a range of pupil background characteristics. The latter were sex (female or not), having a first language other than English (EAL or not), having a special educational need or not (SEN), ethnicity (majority white or ethnic minority), and whether each pupil was eligible for free school meals or not (FSM, a measure of living in a household officially classed as in poverty).

## Exercise 17.2

Pearson's R used for correlation or linear regression can only be conducted with real number variables. Some of the variables listed above are not real numbers, but categories. Which are which? What does this mean for the analysis?

## Dummy variables

Linear regression is only possible with real number variables, which can be treated as a score. However, there is a tradition of also using 'dummy variables'. If a categorical variable is converted to a binary (e.g. 0 or 1) value then the ensuing number code could be treated as a real number representing either 0 or 100% of having a specific category.

In the example above, missing values have been recoded (Chapter 12), and the other categories collapsed so that ethnicity is represented as being white or not, poverty by being FSM-eligible or not, and so on. In the dataset these are coded as values 1 (having the first characteristic) or 2 (not having the first characteristic). These could be accepted as equal-interval 'measures', as discussed in Chapter 2, because they have only one interval, between 1 and 2. In this way, dummy categorical variables are regularly used in linear regression.

This approach goes further when analysts want to preserve more than two categories, and still use them in a regression model. For example, imagine that ethnicity actually had three simple categories – white, Black and Asian in origin. To create the dummy variable above, Black and Asian would be collapsed to one cell, and compared to white only. In the UK, this would contrast the majority with two ethnic minorities combined, but might lose important distinctions between Black and Asian students. An alternative would be to create one dummy binary variable for white or anything else, and another for Black or anything else. Asian origin would then be defined as not white *and* not Black. There should always be one fewer dummies than the number of categories.

Maybe this last idea is a step too far. Maybe the whole thing distorts our data. Can a binary category really be said to cross-plot in a straight line with a real number variable? You have to judge how far you go, but it is best to use dummies sparingly, if at all. We will use them for the following illustration, but in a later section I will explain why they were not needed in the published analysis anyway.

## Running a true multivariate regression

The regression predicting the post-intervention reading score from this real-life study, and using all of the predictor variables above, would be conducted just as before but with many more independent variables.

---

Go to the "Analyze" menu, select "Regression", and from the sub-menu select "Linear". Move PostReading to the Dependent box, and the variables AgeatposttestinMonths, FSM, Sex, SEN, EAL, Ethnicity, PreReading, Attendance.sessions, and Treatmentphase to the Independent(s) box.

```
REGRESSION
/MISSING LISTWISE
/STATISTICS COEFF OUTS R ANOVA
/CRITERIA=PIN(.05) POUT(.10)
/NOORIGIN
/DEPENDENT PostReading
/METHOD=ENTER AgeatposttestinMonths FSM Sex SEN EAL Ethnicity PreReading
Attendance.sessions Treatmentphase.
```

---

Cleaned of all the irrelevant stuff about significance testing, the main output would be like Table 17.5. The overall model correlates with reading scores with an $R$ of 0.69. $R^2$ is 0.47, and in this model with more predictors the adjusted $R^2$ drops to 0.46 (see above). Nevertheless, this is a strongly predictive model with real-life data, accounting for nearly 50% of the variation in reading outcomes.

Table 17.5 Model summary predicting reading score after intervention

| R | R squared | Adjusted R squared |
|---|---|---|
| 0.688 | 0.473 | 0.457 |

The coefficients for each variable are shown in Table 17.6. Unsurprisingly, prior reading score is the best single predictor of later reading score (+0.59). Further than that, the younger the pupil is, the more they gained on average, by a small amount (−0.09), perhaps because they started from a lower base in reading. The more sessions of Switch-on that they attended, the more a pupil in the treatment group gained (the number of sessions for the control group was zero). Apart from all the other factors, those in the intervention group (receiving Switch-on training) do slightly better than others (−0.05 where 1 is intervention and 2 is not). It often takes time to work out what the pluses and minuses on the coefficients mean, and you may have to look back at the coding used in your dataset to remember. There is some indication here that the intervention is beneficial but only weakly (coefficient effect size of 0.05 for treatment group, and 0.06 for number of sessions attended).

**Table 17.6** Coefficients for model predicting reading score after intervention

|  | Unstandardised B coefficient | Standardised beta coefficient |
| --- | --- | --- |
| Constant | 22.75 | – |
| Age at test in months | −0.26 | −0.09 |
| *FSM eligible or not* | 0.95 | +0.05 |
| *Female or not* | −1.32 | −0.07 |
| *SEN or not* | 3.47 | +0.17 |
| *EAL or not* | 6.36 | +0.13 |
| *Ethnicity* | −3.29 | −0.10 |
| Prior reading score | 0.67 | +0.59 |
| Number of sessions attended | 0.03 | +0.06 |
| *Treatment group or not* | −0.85 | −0.05 |

Note: Dummy variables are in italics.

## Forward entry of variables

One way to try and create a simpler model than this is to enter the predictor variables into the model in different ways. So far, the method has been to enter all variables at once. Other alternatives are forward/stepwise (putting in one variable at a time until the next one makes no difference to the model) or backward/remove (entering all variables at once and removing them one at a time until the next one leads to a worse model). Here we will use the same model as above, but with forward as the method, rather than enter.

---

Use the "Analyze" menu, select "Regression", then "Linear" and add the dependent and independent variables as above. Before clicking OK, click the downward arrow in the Method dropdown menu, which by default says Enter (just below the Independent(s) variables). This will offer a choice of Enter (which simply enters all the independent variables at once), Stepwise, Remove, Forward and Backward. Select Forward and click OK.

*(Continued)*

```
DATASET ACTIVATE DataSet1.
REGRESSION
/MISSING LISTWISE
/STATISTICS COEFF OUTS R ANOVA
/CRITERIA=PIN(.05) POUT(.10)
/NOORIGIN
/DEPENDENT PostReading
/METHOD=FORWARD AgeatposttestinMonths FSM Sex SEN EAL Ethnicity PreReading
Attendance.sessions Treatmentphase.
```

The main output looks like this:

**Model Summary**

| Model | R | R Square | Adjusted R Square | Std. Error of the Estimate |
|---|---|---|---|---|
| 1 | .645[a] | .416 | .414 | 7.148 |
| 2 | .663[b] | .440 | .436 | 7.012 |
| 3 | .672[c] | .452 | .446 | 6.949 |

a. Predictors: (Constant), PreReading
b. Predictors: (Constant), PreReading, SEN
c. Predictors: (Constant), PreReading, SEN, Attendance.sessions

**Coefficients[a]**

| Model | | Unstandardized Coefficients B | Std. Error | Standardized Coefficients Beta | t | Sig. |
|---|---|---|---|---|---|---|
| 1 | (Constant) | 23.758 | 3.827 | | 6.207 | .000 |
|   | PreReading | .735 | .050 | .645 | 14.743 | .000 |
| 2 | (Constant) | 23.272 | 3.757 | | 6.194 | .000 |
|   | PreReading | .685 | .051 | .602 | 13.487 | .000 |
|   | SEN | 3.361 | .936 | .160 | 3.590 | .000 |
| 3 | (Constant) | 22.547 | 3.734 | | 6.038 | .000 |
|   | PreReading | .680 | .050 | .597 | 13.500 | .000 |
|   | SEN | 3.480 | .929 | .166 | 3.745 | .000 |
|   | Attendance.sessions | .056 | .022 | .109 | 2.555 | .011 |

a. Dependent Variable: PostReading

There are now three models, with one, two and three predictors in turn, each having a row in the first part of this output. The predictors are selected to be those that are the best at predicting the outcome reading score. No more are needed after model 3, because the fourth best predictor makes so little difference to the result. For each model, the software presents the overall model summary ($R$ and $R^2$), and then the coefficients for any variable in the model at that stage. These figures are cleaned up and presented in Tables 17.7 and 17.8, only for the final model. With an $R$ of 0.67, this far simpler model is nearly as good as the full one in Table 17.5.

**Table 17.7** Model summary predicting reading score after intervention, forward entry

| R | R squared | Adjusted R squared |
|---|---|---|
| 0.672 | 0.452 | 0.446 |

The advantage of this new model is that it is much simpler. Age, FSM, sex, EAL, ethnicity, and treatment group are no longer treated as relevant. Presumably most of these were dealt with by the randomisation, so that they are balanced between the two groups. Modelled in this way, there are only three useful predictors – SEN, prior reading score and the number of sessions. The coefficients for SEN and prior reading are almost exactly the same as in the full model (Table 17.6), which is reassuring. However, the coefficient for number of sessions has doubled. It is likely that this measure has picked up variation from the omitted 'treatment group' variable, since the control group all had zero sessions. This again indicates that the intervention has been effective. In order to see this idea more clearly, the next section illustrates regression with more than one block of predictors.

**Table 17.8** Coefficients for model predicting reading score after intervention, forward entry

|  | Unstandardised B coefficient | Standardised beta coefficient |
|---|---|---|
| Constant | 22.55 | – |
| SEN or not | 3.48 | 0.17 |
| Prior reading score | 0.68 | 0.60 |
| Number of sessions attended | 0.06 | 0.11 |

In summary, though, the final model here is almost as predictive as the full model, but has only three predictors rather than nine. This makes it a much better model.

## Biographical and other models

Apart from SEN, none of the pupil background characteristics were retained in the model above. Does this mean that they are not important? Also in the model above there was no room for the treatment group variable. Does this mean that the treatment group did not matter, or was it overwhelmed by knowledge of the number of sessions? In order to help understand which variables might be **proxy** or substitute measures for others we can create the model more carefully.

In this example, we will enter the variables into the model in stages representing a time sequence (biographically). We will put all of the background variables (like FSM) into the first block as usual (using forward entry again). Then we will create a second block, and put prior reading score in that. In this way, we can look at the background variables in isolation, and then prior attainment net of the influence of background. The justification is theoretical, given that background comes before early reading ability, biographically. Sex, EAL, poverty and so on might well influence reading score at the start of the term aged 11, but 11-year-old reading ability cannot influence ethnicity, childhood poverty and so on.

In a third block we will add the treatment group (dummy), and in a fourth block we will add the number of treatment sessions attended. This means we can see if there is any evidence that the treatment was effective, net of background and prior attainment, without being confused by the number of sessions. We can then address the separate question of whether the number of sessions attended might make a difference, over and above simply being in the treatment group.

---

From the "Analyze" menu, select "Regression" and then "Linear". Move the variables AgeatposttestinMonths, FSM, Sex, SEN, EAL, and Ethnicity to the Independent(s) box. Select Forward from the Method dropdown menu. The area around the Independent(s) box is titled Block 1 of 1. Click Next, just below the title. This creates an empty Independent(s) box again but for Block 2 of 2. You can click Previous to go back to Block 1, and Next to go forward again. Add the variable PreReading as a predictor for Block 2, click Next (it does not matter which method is used because there is only one variable in this block), add Treatmentphase for Block 3, click Next, add Attendance.sessions for Block 4, then click OK.

REGRESSION

/MISSING LISTWISE

/STATISTICS COEFF OUTS R ANOVA

/CRITERIA=PIN(.05) POUT(.10)

/NOORIGIN

/DEPENDENT PostReading

/METHOD=FORWARD AgeatposttestinMonths FSM Sex SEN EAL Ethnicity

/METHOD=ENTER PreReading

/METHOD=ENTER Treatmentphase

/METHOD=ENTER Attendance.sessions.

The output looks like this:

**Model Summary**

| Model | R | R Square | Adjusted R Square | Std. Error of the Estimate |
|---|---|---|---|---|
| 1 | .324a | .105 | .102 | 8.851 |
| 2 | .345b | .119 | .113 | 8.796 |
| 3 | .667c | .444 | .439 | 6.996 |
| 4 | .676d | .457 | .450 | 6.928 |
| 5 | .676e | .457 | .448 | 6.938 |

a. Predictors: (Constant), SEN
b. Predictors: (Constant), SEN, Sex
c. Predictors: (Constant), SEN, Sex, PreReading
d. Predictors: (Constant), SEN, Sex, PreReading, Treatmentphase
e. Predictors: (Constant), SEN, Sex, PreReading, Treatmentphase, Attendance.sessions

**Coefficients**a

| Model | | Unstandardized Coefficients B | Std. Error | Standardized Coefficients Beta | t | Sig. |
|---|---|---|---|---|---|---|
| 1 | (Constant) | 71.234 | 1.531 | | 46.541 | .000 |
| | SEN | 6.792 | 1.137 | .324 | 5.972 | .000 |
| 2 | (Constant) | 74.991 | 2.288 | | 32.775 | .000 |
| | SEN | 6.630 | 1.133 | .316 | 5.853 | .000 |
| | Sex | -2.243 | 1.020 | -.119 | -2.198 | .029 |
| 3 | (Constant) | 25.904 | 4.109 | | 6.305 | .000 |
| | SEN | 3.304 | .935 | .157 | 3.535 | .000 |
| | Sex | -1.274 | .815 | -.067 | -1.564 | .119 |
| | PreReading | .678 | .051 | .595 | 13.326 | .000 |
| 4 | (Constant) | 29.477 | 4.289 | | 6.873 | .000 |
| | SEN | 3.380 | .926 | .161 | 3.649 | .000 |
| | Sex | -1.368 | .808 | -.072 | -1.694 | .091 |
| | PreReading | .673 | .050 | .591 | 13.344 | .000 |
| | Treatmentphase | -2.088 | .792 | -.112 | -2.635 | .009 |
| 5 | (Constant) | 27.785 | 6.089 | | 4.563 | .000 |
| | SEN | 3.400 | .929 | .162 | 3.660 | .000 |
| | Sex | -1.355 | .810 | -.072 | -1.673 | .095 |
| | PreReading | .673 | .051 | .591 | 13.319 | .000 |
| | Treatmentphase | -1.256 | 2.266 | -.067 | -.554 | .580 |
| | Attendance.sessions | .025 | .063 | .048 | .392 | .695 |

a. Dependent Variable: PostReading

Now there are five steps in the model. There are two versions for Block 1, because two predictors are entered Forward. These are SEN as before, and now also Sex. After these, the other background variables are irrelevant. SEN (with R of 0.324) contributes about half of the value of R for the whole model with all variables included. Sex adds only a little more to R (to 0.345). Then there is one version each of Blocks 2, 3 and 4 (steps 3, 4 and 5 in the output). Prior attainment explains about half again (0.345 to 0.667). Then being in the treatment group adds a little more (0.667 to 0.676). Perhaps most interestingly, the final step (5) makes now no difference to the R-value in the overall model.

Cleaned up as usual, the results for the last step are as in Tables 17.9 and 17.10. R is 0.68, again a pretty strong explanatory model.

**Table 17.9** Model summary predicting reading score after intervention, biographical

| R | R squared | Adjusted R squared |
|---|---|---|
| 0.676 | 0.457 | 0.448 |

The coefficients confirm this. Again prior reading is the strongest predictor, and the value for SEN is also about the same as in the other models. This is a sign of a solid, stable predictor. Males do only slightly worse on average (–0.07), and the intervention group do slightly better (–0.07, where treatment is coded as 1 and control as 2). The number of sessions is now the weakest predictor.

**Table 17.10** Coefficients for model predicting reading score after intervention, biographical

| | Unstandardised B coefficient | Standardised beta coefficient |
|---|---|---|
| Constant | 27.79 | – |
| SEN or not | 3.40 | +0.16 |
| Sex | –1.36 | –0.07 |
| Prior reading score | 0.67 | +0.59 |
| Treatment or not | –1.26 | –0.07 |
| Number of sessions attended | 0.03 | +0.05 |

The R for step 5 is the same as for step 4, which suggests that the number of sessions is actually irrelevant over and above the knowledge of which treatment group a pupil is in. Therefore the model for step 4 should be preferred. It involves fewer predictors (**parsimony**) and has a slightly larger adjusted $R^2$ (0.45). It is a better model (based on a judgement). Almost everything stays the same with one variable fewer, and the impact of the treatment is clearer (–0.11). Even if SEN, sex, and prior reading score were not exactly balanced between the treatment and control groups in this experiment, we have evidence that the treatment was helpful (Table 17.11). We are in a better position to make a causal claim here than we are usually with a regression result, because the data is from an experimental design.

**Table 17.11** Coefficients for most parsimonious model predicting reading score after intervention

|  | Unstandardised B coefficient | Standardised beta coefficient |
| --- | --- | --- |
| Constant | 29.48 | – |
| SEN or not | 3.40 | +0.16 |
| Sex | –1.37 | –0.07 |
| Prior reading score | 0.67 | +0.59 |
| Treatment or not | –2.09 | –0.11 |

It may seem odd to have so many slightly different ways of modelling regression analysis with the same data. But there are many more, using interactions or not, as many blocks as you want, and different methods for entering the variables within each block. The precise version that you use might be justified by your research question (see the example in Chapter 19), or by theoretical considerations, as in the example above, or by some other factor. Whatever your reasoning you must explain and justify this to your reader. There is no such thing as a *standard* multiple linear regression.

For the Switch-on example, the best model is probably the last one, considering the possible impact of the intervention net of pupil background and prior attainment, and then seeing if any further variation can be explained by the frequency of teaching sessions in the intervention group. However, in reality, I did not do any of these analyses (Gorard et al., 2015). Even the complex four-step regression model cannot be used by itself to test whether the Switch-on intervention *caused* the relative improvement of one group over the other.

The warrant for claiming that the Switch-on intervention had an impact stems solely from the appropriate experimental design for that research question (Chapter 9). I would not suggest a regression analysis for experimental data, under normal circumstances. The experimental design does the hard work, and requires only the simplest analysis – a comparison of the reading scores in each group before the intervention to check for imbalance, and a comparison of the reading scores in each group after the intervention to check for differential improvement by one group, then creating an effect size to summarise the finding. Note that this simpler analysis also solves the problem of whether to treat dummy variables as real numbers. There is no place for these variables in the usual effect size calculation. As long as the scale is sufficient we can rely on the randomisation.

The forward, backward and similar methods for entering variables in each block should be based on either theoretical or substantive contribution to the model. Always check that your software is not using hidden significance tests to make these choices. In case they are it is usually safer for you to retain control of the process.

If you want to use a forward approach just enter the variable with the highest (*R*) correlation with the outcome, then add all others individually with that first variable, and see which of them adds most to the overall *R*, and so on. This sounds a bit fiddly but it not only retains control but also gives you a better idea of how the model is being built up.

### Exercise 17.3

Find an article in your area of interest that uses regression. Prepare a critique, noting how well and fully the paper presents the methods, and whether the paper includes undigested computer output, or whether the tables are made easy to read. Does the paper use causal words like 'influence' or 'impact' without justification, or without making it clear that it is speculating?

## CONCLUSION

In several forms of multivariate analysis, the order in which independent variables are entered into the explanatory model can make a very substantial difference to the results obtained. Different explanations of social phenomena can be derived using the same technique but with relatively minor variations in the order of entering variables. Since many well-known and influential theories are based on precisely such models, the importance of bearing this principle in mind is difficult to exaggerate. Put simply, in the absence of greater detail about the order in which variables are considered, some of these relatively well-known theories may be less secure than previously imagined (Gorard and Rees, 2002).

Multiple regression is an easy-to-use, and powerful, method for summarising any patterns in large datasets. It is also an entry into the world of modelling and complex multivariate analyses. However, regression does not really 'predict' or even explain anything. It is not generally successful when used to predict real-life events such as stock market crashes or thunderstorms (Brighton, 2000). Regression can do the same as a graph can, but with more variables at once than could be conveyed in a readable graph. It can encourage you to consider the interactions between those variables, and identify possible linkages. But like a graph, regression does not really tell you anything new, and it is not a *test* of any model. It is a way of helping you to think about the meaning of your results. As such, it is more like the start of any investigation, trying to explain what those linkages are, rather than the final result (Phillips, 2014).

In addition to the ideas in this chapter, there are correlation and regression techniques available for use with categorical variables (Siegel, 1956). The next chapter looks in more detail at the assumptions underlying a linear regression analysis. Chapter 19 then looks at an approach to regression that involves fewer assumptions and is much more suitable for working with our more usual numeric datasets, which contain both real numbers *and* categorical variables.

## Notes on selected exercises

### Exercise 17.1

Adding a second predictor might increase the value of R (whether positively or negatively), even if only slightly. The final R with two predictors will not usually be the sum of the R-values from the regression (or correlation) involving each predictor and the independent variable. This is because the predictors may be intercorrelated themselves, and so act as overlapping predictors to some extent. For example, predicting people's income from their level of education and their parents' income may yield an R of 0.6 from regression. A regression of individual income and level of education might yield an R of 0.3, and a regression of individual income and parental income might yield an R of 0.5. Putting both together does not create a regression model with an R of 0.8 (the sum of 0.3 and 0.5) because level of education and parental education are also correlated to some extent (perhaps with an R of at least 0.2).

### Exercise 17.2

As discussed in the text, the variables like SEN, FSM and Sex are not real numbers, but dummies. Dummies are often used in regression. Whether this should be so is not clear. Ideally, categorical variables should be respected for what they are, and used with an approach that does this – such as logistic regression, described in Chapter 19.

### Exercise 17.3

The publication of large, unwieldy and unhelpful tables is not an issue only for regression, but is perhaps at its worst when regression of any type is used. Indications that the author does not care how well their reader understands the analysis also include the use of abbreviated, or computer-coded, names for variables, lots of asterisks and footnotes, too many decimal places, and having more than one number per table cell (such as when both mean and standard deviation appear in one cell). For more on this see Chapter 21. But there are many more abuses. Even more important is the free use of causal terms like 'influence' and 'impact'. It is perfectly proper to run a regression model and then speculate on whether the pattern uncovered is a causal one. However, the reader must be clear that the ideas are speculative. But what largely happens is that authors claim a causal relationship based just on regression (or close relative of regression like path analysis).

## Further exercises to try

Find a suitable real-life dataset you are interested in (or use one provided on the website). Select a real number variable as the outcome, and use other variables as predictors in a regression analysis. Make sure you only use real number (or dummy variable) predictors. Try entering the predictors in two blocks. What differences does this make?

## Suggestions for further reading

A chapter introducing some basic ideas on regression:
Gorard, S. (2021) Multiple linear regression. In R. Coe, L. Hedges, M. Waring and L. Day-Ashley (eds), *Research Methodologies and Methods in Education*, 3rd edition. London: Sage.

A fuller and appropriately balanced account of regression:
Berk, R. (2004) *Regression Analysis: A Constructive Critique*. Thousand Oaks, CA: Sage.

An open access paper using linear regression, based on the techniques in this chapter, and having important implications for policy:
Gorard, S. and Siddiqui, N. (2019) How trajectories of disadvantage help explain school attainment, *SAGE Open*, 9(1). https://journals.sagepub.com/doi/10.1177/2158244018825171

# 18
# ASSUMPTIONS AND LIMITATIONS IN REGRESSION

## SUMMARY

This brief chapter is the last of the more technical and conceptual chapters in this book, and it could be skipped by readers who want to move on to reading about logistic regression and other forms of multivariate analysis. The chapter expands on the important assumptions underlying correlation and regression, as outlined in Chapters 16 and 17. And it discusses why most of us will never need to use the more complex versions of linear regression now quite widespread in the literature.

## INTRODUCTION

Chapter 15 introduced the idea of correlation and the Pearson's $R$ correlation coefficient. Chapters 16 and 17 extended this to simple and multiple regression, based on such correlations. Each chapter also described the key assumptions underlying these techniques – the use of only real number variables, and an approximate linear relationship between each predictor variable and the predicted variable. This chapter looks at some additional and more complex assumptions underlying regression analysis, which you may encounter in your reading. This makes the chapter more technical than the previous one, and less important for most readers because these additional assumptions either are irrelevant in most contexts, or make little difference in practice. The chapter could be skipped by new researchers.

## LIMITATIONS OF MULTIPLE LINEAR REGRESSION

Linear regression is frequently used in modelling situations for which it was not intended. These include those common situations in social science where some of the independent variables are categorical in nature. We often want to look at patterns in terms of sex, gender, ethnicity, social class, and so on.

Standard regression does not work well with such categorical independent variables, especially those having more than two values (Hagenaars, 1990). If a variable has only two possible values then it could be treated as an equal-interval dummy variable (since there is only one interval), as described in Chapter 17. Further, even variables with more than two categories might be used by converting them to a series of dummy variables. A social class scale with three categories, such as service, intermediate and working class, could be treated as two dummy variables. The first dummy is a yes/no variable representing being in the service class or not, and the second dummy represents being in the intermediate class or not. Working class is therefore defined as being not service and not intermediate class.

However, some writers have argued that this treatment of categories is a distortion and not really appropriate, especially now that newer methods have been developed specifically to deal with categorical variables (see Chapter 19). It is sometimes assumed in regression that the variables are normally distributed (Lee et al., 1989), but dummy variables cannot have such a distribution. Simply converting a categorical variable into a set of dummies is not the full solution. Using dummy variables may even add to the measurement error (Blalock, 1964).

Another problem is that because regression models are fitted to the data after collection (i.e. they are not really 'predictions'), it is possible for the analytical software to use the natural but often meaningless variation in the independent variables to match individual scores in the purportedly dependent variables. Any model that results is then partly or completely irrelevant, but can have a very high $R$-value. For example, a dataset with as many independent variables as cases will always yield an $R$-value of 1, even if the scores are all randomly generated (Gorard, 2006a). This is partly why reputable texts emphasise that the number of cases in any study must outnumber the number of variables by at least an order of magnitude.

If we vary the analysis to use backward elimination of any redundant variables , it is possible to reduce the number of independent variables in a fully random number model without substantially reducing the $R^2$ value. In other words, we can still create a perfect prediction/explanation for the dependent variable, but this time using fewer variables than cases. If we are happy to allow the $R^2$ value to dip below 1.0 then the number of variables needed to predict the values of the dependent variable can be reduced dramatically. It is easily possible, in this way, to produce a model with an $R$-value of 0.6 or higher using only 10 or fewer variables for 100 cases, even where all of the values are nonsensical random numbers. This $R$-value is higher than many of those that are published in journals and that have been allowed to affect policy or practice. And the ratio of cases to variables would appear to be 10:1, which can be considered reasonably healthy.

This is part of the reason for being very cautious about small values of $R^2$, and for checking the adjusted $R^2$ (Chapter 17). Unfortunately, there is no standard scale of substantive importance for effect sizes like $R^2$ (Gorard, 2006b). You have to judge the meaning for yourself (Chapter 8).

## ASSUMPTIONS OF MULTIPLE LINEAR REGRESSION

The basic assumptions for running a regression analysis, other than the two discussed so far, are as follows:

- All variables should be measured as far as possible without error.
- No independent variable should have a perfect linear correlation with another (known as perfect 'multicollinearity').

If the four basic assumptions do not hold, then the regression results can be misleading (Maxwell, 1977; Menard, 1995).

There are many more, but less important, assumptions. The full list of these can be bewildering as well as unrealistic. They include the following:

- The measurements are from a random sample (or at least a probability-based one).
- There are no extreme outliers (scores apparently of a different order of magnitude than the rest).
- The dependent variable is approximately normally distributed (or at least the next assumption is true).
- The **residuals** for the dependent variable (the differences between calculated and observed scores in the regression) are approximately normally distributed.
- The variance of each variable is consistent across the range of values for all other variables (or at least the next assumption is true).
- The residuals for the dependent variable at each value of the independent variables have equal and constant variance.
- The residuals are not correlated with the independent variables.
- The residuals for the dependent variable at each value of the independent variables have a mean of zero (or they are approximately linearly related to the dependent variable).
- For any two cases the correlation between the residuals should be zero (each case is independent of the others).

However, these lesser assumptions are disputed in terms of what they mean, and of the implications of running an analysis that does not meet them (de Vaus, 2002; Miles and Shevlin, 2001). Most obviously, many of the assumptions are irrelevant if you are not planning to use significance tests or similar (Chapter 6). This book proposes the modern and logical approach of not using significance tests. Therefore, assumptions like the first one in this list are not relevant. Regression analyses are perfectly possible with population figures, convenience samples and incomplete random samples. Indeed, working with population figures is recommended.

Some of the other assumptions are only relevant to the constant in the regression model (Berry and Feldman, 1985). Some might affect the coefficients, but only slightly. In general, if any of these lesser assumptions are not true for any analysis, the impact is to reduce the apparent size of any relationship uncovered. Therefore, and in general, if you obtain a powerful result it is still relatively safe to proceed to investigate it. Remember, regression is a form of modelling and not a definitive test of anything.

## THE INDEPENDENCE OF CASES AND HIERARCHICAL CLUSTERS

The final assumption above, that all cases used in a regression model are independent of each other, also arose when looking at random sampling in Chapters 10 and 11. In a true random sample, all cases would be independent of each other.

This assumption is sometimes not true in practice, in several ways. Perhaps most simply, it is not true in a longitudinal design, where measures are taken repeatedly over time from the same cases (Chapter 9). Such data could be analysed as repeated cross-sectional data, so that no case appears more than once in each analysis. This is quite common with experimental data taken both before and after an intervention.

However, sometimes repeated measures are routinely used in the same analysis in practice, such as in studies of school effectiveness. This is done without creating major problems, because the independence assumption is really only about influencing the standard errors in an imagined significance test. If you do not plan to conduct significance tests, as you should not, this assumption is one to be aware of in your design, but it is not crucial in your choice of analysis.

Another way in which cases might appear not completely independent of each other is that cases with certain characteristics are somewhat more likely to be like other cases with the same characteristics than like cases without them. As an example, all close members of one family are more likely to report the same ethnicity than members of different families do. This is how society currently is. And it does not mean that we cannot use a regression model, with variables representing family or ethnicity. In some important medical respects male patients in all hospitals may be more alike than the male and female patients in one hospital are. A random sample is still a genuine random sample even if some cases attend the same church, or buy their clothes from the same shop. These illustrations are made to show that cases can be alike or not alike, in families or across organisations, and we routinely and correctly ignore this as an assumption for our sampling, and our analyses. Of course, we may want to investigate these patterns and see who uses particular shops, for example, but this does not flout the assumption of independence.

**Image 18.1**  Hierarchies everywhere

However, you may see some researchers and some methods resources worry about the non-independence of cases much more than they should, because they are still using significance tests with cluster randomised samples. But they only worry about similarities between cases that nest in hierarchical clusters (wards within hospitals), and ignore the majority of similarities like those above that do not nest (sex and ethnicity). It is possible that cases within clusters are more similar to each other in some respects than they would be to cases in the population more generally. If so, according to some researchers this can lead to what is called a 'design effect' that is claimed to reduce artificially the standard error of a sample, leading to the possibility of a misleading result from a significance test analysis (or confidence interval or power analysis).

'Multi-level modelling' (MLM) or 'hierarchical linear modelling' are names for a range of techniques you may come across, developed from hierarchical approaches to analysis in fields such as agriculture (Aitkin and Longford, 1986; Raudenbush and Bryk, 1986). They are an attempt to overcome the supposed problems created by flouting the assumption of independence of cases.

We might expect people in one family to be more like each other than like a person in another family. We might expect children in one school to perform more like each other than a child in another school. Such cases are said to be **autocorrelated** in the same way that repeated measures of the same cases can be. MLM advocates argue that since this is the case, we are better off building these similarities into our analytical methods, when we use cluster samples (Chapter 10).

MLM is therefore a special form of regression that allows the analyst to use both individuals and groups of individuals in the same model but to avoid flouting the last assumption (of independent cases), since the standard error of any results can be affected by the clustered nature of the data.

However, this whole idea can be seen as contradictory. If the assumption for the use of regression techniques – that measurements are taken from a true random sample – is met, then MLM is not needed. The cases will be independent by definition. Any apparent correlation between cases that exists in the sample will also be random (accidental) rather than the result of social structures. For example, if the population is the pupils in all schools in England, then a random sample will have a national spread. There may be more than one pupil from the same school, just as there will almost certainly be more than one with the same age, sex, prior attainment, class and ethnic background.

It may be important that all of these variables, and more, are taken into account in the regression. But this is very different from the argument for MLM which is based on a very different kind of sampling, which is not truly random. It is only concerned with hierarchical structures, and so it actually ignores similarities that do not nest within clusters. Those ignored include key variables such as sex, ethnicity, occupation and the majority of important variables that interest social scientists, but do not form a clear hierarchy.

If the sample is based on randomised clusters, then the analysis can properly be done at the level of the clusters, to avoid an ecological fallacy and other dangers (Moses, 2001). This is what the British Medical Association, Cochrane Collaboration and other authorities suggest. A good cluster random sample mimics a true random sample (as might a systematic or stratified one), and in so far as it does it then allows an individual-level analysis without concern for autocorrelation at the cluster level. Put simply, autocorrelation is a deficiency of sampling and not of analysis, and the appropriate solution is therefore better sampling and not more complex analyses.

On the other hand, if the assumption that the sample is random is not met then there will be no standard error in the cases (by definition). No significance tests can be performed, and so the results cannot be influenced by an underestimate of the purported standard error.

In summary, whether your sample is clustered or not you will *always* get the same effect sizes in your results whether you use simple regression or something more complex. And the design effect from clustering is anyway often substantively zero in practice (Xiao et al., 2016). Fitz-Gibbon (2001) analysed five large datasets using both standard regression and MLM. The two sets of results from each analysis correlated at around 0.99 (Fitz-Gibbon, 2001). In this case, the complex results based on MLM were less accessible to a wider readership for no good reason. They are not as parsimonious as those from traditional regression (requiring more parameters to be estimated), they are less generalisable (i.e. more specific to the dataset they are fitted to), need a larger dataset, and are more complex to estimate.

We do not have the statistical techniques to adjust our deficient data appropriately (Moses, 2001). Greater complexity of an analysis is not the answer; better data is

(Brighton, 2000). To increase robustness, it is simpler just to increase the sample size (Raudenbush, 2002), or use a number of other simple adjustments or robust alternatives (Bland, 2003; Gorard, 2007).

## CONCLUSION

This chapter has looked at some of the broader assumptions you may see listed as underlying regression models in your wider reading. It bears repeating that most of these assumptions are irrelevant to us because we are not using significance tests. Most of the rest of the assumptions will not make a substantial difference to the results anyway. Your outcome and predictor variables must be real numbers, and they must cross-plot to a near-straight line. In addition, and as in all research, the errors in your measurements should be minimised. And finally, it is safer not to use predictor variables that are themselves perfectly intercorrelated. The approaches described in Chapter 17 are sufficient for most real-life research.

### Suggestion for further reading

A short book discussing the assumptions of regression, for those who want to pursue this:
Berry, W. and Feldman, S. (1985) *Multiple Regression in Practice*. London: Sage.

# 19
# PREDICTING OUTCOMES USING LOGISTIC REGRESSION

## SUMMARY

This chapter looks at examples of using logistic (sometimes called logit) regression, focusing on a predicted outcome or dependent variable with two outcomes, and as many predictor or independent variables as required. Logistic regression is easier and more flexible than linear regression in terms of its assumptions and its more natural use of categorical predictors (Chapter 17). However, it has the limitation of only being able to model a categorical outcome (not a real number variable). It could be used to address research questions like:

- What are the strongest predictors of getting or not getting a job?
- Net of other factors, does attendance at a private school increase the chances of attending university?
- What are the key characteristics of successful election candidates?

## LOGISTIC REGRESSION

Linear regression is a useful and powerful way of modelling data for some situations (Chapter 17). But it can only model a real number outcome, is not elegant or even appropriate for use with categorical predictor variables, and it has underlying assumptions such as linearity that might not be met in real-life situations (Brignell, 2000). Logistic regression can cope with both real number and categorical predictors in a natural way, and has the added advantage that the dependent and independent variables need not be linearly related. Perhaps because of this, it has been claimed that logistic regression can routinely explain more of the variance in the model than linear regression models or other analyses, using the same dataset.

Binary logistic regression uses predictor variables (of any sort) to compute a probability score for a predicted version of the dependent (categorical) variable. If this score is above a specified critical value (default 0.5), then the predicted dependent variable is set to one category, else it is set to the other. In other words, the procedure is used to 'predict' which of two categories each individual case will manifest, and in doing so creates a model based on the predictor variables (Gilbert, 1993; Lehtonen and Pahkinen, 1995). The model is called 'logistic' because the calculations are based on the logarithm of the odds of being in one of the two outcome categories. The model, therefore, produces an overall strength of the prediction (a bit like $R$ for linear regression), and the log odds for being in one category based on each variable (a bit like the unstandardised coefficients in linear regression). These log odds can be converted back to plain odds for easier understanding (a bit like the standardised coefficients in linear regression).

## A WORKED EXAMPLE

This is best explained in practice with a simple example. Imagine that a dataset of 100 cases includes three variables – a category of being in employment or not, an ordinal category of level of education (rising from compulsory, to post-compulsory, to degree level or above), and a real number (age in years). We can create a model that would try to predict/explain an individual's employment category based on their age and education. As with linear regression, it would be best to do some simpler descriptive analyses first, to clean and get to know your dataset and variables before conducting a logistic regression (Chapter 3).

Regardless of how the binary outcome variable is coded in the dataset (e.g. 'employed' or 'unemployed', '1' or '2', 'A' or 'B') the model will always recode it as '0' and '1'. You need to check when this happens that you remember which category is which – you can easily arrange the coding so that '1' means employed, but you cannot just assume that it will be so.

Similarly, regardless of how the predictor categorical variables are coded in your dataset, the model may recode them (usually in the order that the categories are listed in). And again you need to note how this coding is done so that there is no confusion. This will become clearer as we work through the example.

---

To conduct a logistic regression in SPSS, go to the "Analyze" menu, select "Regression" and then "Binary Logistic". Move the variable Employed to the Dependent variable box, and Age and Qualification to the Covariates (independent variable) box. You need to tell the model which variables

*(Continued)*

are to be treated as categorical. Go to the top right of the dialogue screen, click "Categorical", and move Qualification from the Covariates to the Categorical Covariates box. Click OK.

```
LOGISTIC REGRESSION VARIABLES Employed
  /METHOD=ENTER Age Qualification
  /CONTRAST (Qualification)=Indicator
  /CRITERIA=PIN(.05) POUT(.10) ITERATE(20) CUT(.5).
```

The output, which is surprisingly long and rather confusing at first, will look something like this:

| Logistic Regression | | | |
|---|---|---|---|
| Case Processing Summary | | | |
| Unweighted Cases[a] | | N | Percent |
| Selected Cases | Included in Analysis | 100 | 100.0 |
| | Missing Cases | 0 | .0 |
| | Total | 100 | 100.0 |
| Unselected Cases | | 0 | .0 |
| Total | | 100 | 100.0 |
| a. If weight is in effect, see classification table for the total number of cases. | | | |

| Dependent Variable Encoding | |
|---|---|
| Original Value | Internal Value |
| Unemployed | 0 |
| Employed | 1 |

| Categorical Variables Codings | | | | |
|---|---|---|---|---|
| | | | Parameter coding | |
| | | Frequency | (1) | (2) |
| Qualification | Compulsory education | 25 | 1.000 | .000 |
| | Post-compulsory education | 51 | .000 | 1.000 |
| | Degree or higher | 24 | .000 | .000 |

Block 0: Beginning Block

| Classification Table[a,b] | | | | | |
|---|---|---|---|---|---|
| | | | Predicted | | |
| | | | Employed | | Percentage Correct |
| | Observed | | Unemployed | Employed | |
| Step 0 | Employed | Unemployed | 0 | 41 | .0 |
| | | Employed | 0 | 59 | 100.0 |
| | Overall Percentage | | | | 59.0 |
| a. Constant is included in the model. | | | | | |
| b. The cut value is .500 | | | | | |

## PREDICTING OUTCOMES USING LOGISTIC REGRESSION

**Variables in the Equation**

| | | B | S.E. | Wald | df | Sig. | Exp(B) |
|---|---|---|---|---|---|---|---|
| Step 0 | Constant | .364 | .203 | 3.204 | 1 | .073 | 1.439 |

**Variables not in the Equation**

| | | | Score | df | Sig. |
|---|---|---|---|---|---|
| Step 0 | Variables | Age | .423 | 1 | .516 |
| | | Qualification | 13.622 | 2 | .001 |
| | | Qualification(1) | 7.290 | 1 | .007 |
| | | Qualification(2) | .197 | 1 | .658 |
| | Overall Statistics | | 13.622 | 3 | .003 |

Block 1: Method = Enter

**Omnibus Tests of Model Coefficients**

| | | Chi-square | df | Sig. |
|---|---|---|---|---|
| Step 1 | Step | 14.879 | 3 | .002 |
| | Block | 14.879 | 3 | .002 |
| | Model | 14.879 | 3 | .002 |

**Model Summary**

| Step | -2 Log likelihood | Cox & Snell R Square | Nagelkerke R Square |
|---|---|---|---|
| 1 | 120.493[a] | .138 | .186 |

a. Estimation terminated at iteration number 5 because parameter estimates changed by less than .001.

**Classification Table[a]**

| | Observed | | Predicted | | |
|---|---|---|---|---|---|
| | | | Employed | | Percentage Correct |
| | | | Unemployed | Employed | |
| Step 1 | Employed | Unemployed | 16 | 25 | 39.0 |
| | | Employed | 9 | 50 | 84.7 |
| | Overall Percentage | | | | 66.0 |

a. The cut value is .500

**Variables in the Equation**

| | | B | S.E. | Wald | df | Sig. | Exp(B) |
|---|---|---|---|---|---|---|---|
| Step 1[a] | Age | .000 | .013 | .000 | 1 | .992 | 1.000 |
| | Qualification | | | 11.234 | 2 | .004 | |
| | Qualification(1) | -2.522 | .753 | 11.217 | 1 | .001 | .080 |
| | Qualification(2) | -1.670 | .682 | 5.995 | 1 | .014 | .188 |
| | Constant | 1.939 | .864 | 5.041 | 1 | .025 | 6.955 |

a. Variable(s) entered on step 1: Age, Qualification.

The case processing summary (the first bit) tells us how many valid cases there are. There are 100 cases, and none are missing. The dependent variable encoding tells us that the 0 outcome in the model will represent unemployed, and that 1 will be employed status. This is fine. The categorical variables codings are reminding us that where variables have more than two categories (as with Qualification), they will be coded as a series of binary values. The first category is compulsory education, compared to anything else. The second is post-compulsory compared to anything else. The third is implied (it is simply neither of the first two categories), and would represent having a degree compared to anything else. This is also fine.

We can ignore the omnibus test of model coefficients, which involves significance testing and has no relevance for us (Chapter 6). The model summary contains a pseudo-$R$-value, comparable to the $R$ used in linear regression (Chapter 17). Some people use it, but the classification table that follows is a more precise and comprehensible picture of what the model is doing ('pseudo' means false or pretend).

Ignore the model summary (at least for the moment), and the significance clutter as always. This reduces the output to two blocks (labelled 0 and 1) each consisting of a classification table, and a table of variables in the equation (and/or of variables not in the equation). These have been edited in Word to help simplify the output (Tables 19.1–19.3). In recreating these tables we can ignore any columns for S.E. (the standard error), Wald, df (degrees of freedom), and Sig. (significance). The 'cut value' (here 0.5) means that the model predicts any case with a probability of 50% or more as being employed, otherwise unemployed (see below for more on this).

Block 0 for any logistic regression is always the model without any predictors. As such, it simply lays out the number of cases that are actually (observed to be) in each outcome group. Table 19.1 shows us that, of our 100 cases, 59 are observed to be employed and 41 are unemployed. Of course, in a real analysis we should already know this, as we would have run frequency reports and similar (Chapter 3). At this stage (Block 0), if we wanted to predict which outcome category any case would be in, then it would be a pure guess. Our best guess for any case would be 'employed', because this would be correct 59% of the time. So all cases are listed in Table 19.1 as 'predicted employed' and the model summary is 59% correct overall. At this stage there are no variables in the model, and so the two reports after the 'classification table' in Block 0 can be ignored.

**Table 19.1** Base model from logistic regression predicting employment

|  | **Predicted unemployed** | **Predicted employed** | **Percentage correct** |
|---|---|---|---|
| Observed unemployed | 0 | 41 | 0 |
| Observed employed | 0 | 59 | 100 |
| Overall correct |  |  | 59 |

Block 1 is the next step in the model, adding all of the predictor variables in one step. Table 19.2 repeats the useful information from the classification table for Block 1. There are still 41 cases in the row labelled 'observed unemployed' and 59 in the row for 'observed employed'. This will remain constant whatever happens in the model. But now the predictor variables have improved our ability to predict the outcome for each case, from 59% to 66%. This is not a great improvement (remember, this is made-up data, and the model has only two predictors), but it is an improvement. The model predictions are still skewed towards the majority outcome of employed.

**Table 19.2** Full model from simple logistic regression predicting employment

|  | Predicted unemployed | Predicted employed | Percentage correct |
|---|---|---|---|
| Observed unemployed | 16 | 25 | 39.0 |
| Observed employed | 9 | 50 | 84.7 |
| Overall correct |  |  | 66.0 |

By increasing the correct prediction of being unemployed from 0 (Table 19.1) to 39% (Table 19.2) we have lost a little of our accuracy in predicting being employed (dropping from 100% to 85%). This is normal. If we add more variables we should be able to improve the 66% correct, and so have more of the cases in the leading diagonal of Table 19.2 (moving towards an ideal of 100% correct, with more like 41 cases in the first cell and 59 in the fourth cell). What we want from a powerful logistic regression model is to be able to discriminate between the two outcomes safely.

## Exercise 19.1

If our model predicted perfectly, how many cases would there be in the second cell of the first row (which has the value 25 in Table 19.2)?

Before examining how we might improve this model and its discriminatory power, it is best to complete our consideration of the output. Table 19.3 reproduces the 'variables in the equation' section from Block 1, and this shows the odds or coefficients for each variable used as a predictor of employment status. We will ignore the 'constant' for the moment. The column labelled $B$ shows the logarithm of the odds, so it is usually easier to work with the column labelled Exp($B$). This is the antilogarithm of the log odds, or more simply just the odds.

The result for the real number variable Age is perhaps the simplest to understand. The Exp($B$) coefficient represents the odds by which the outcome of being employed increase or decrease with every year of age. Here, however, the odds are exactly 1. If we multiply the chances of being employed by 1 this makes no difference, and if we do this many times for however many years old the person is, the chances will not change. This mean that running the model without age would yield the same percentage correct (66%), and that, for this imaginary model, age does not matter once qualification is accounted for. We will look at a different example later in which a real number variable does seem to make a difference.

Table 19.3  Coefficients for variables in a simple regression model predicting employment

|  | B | Exp(B), or odds |
| --- | --- | --- |
| Age | 0 | 1 |
| Compulsory education | −2.52 | 0.08 |
| Post-compulsory education | −1.67 | 0.19 |
| Degree or higher | - | - |

With a real number variable the odds are multiplied by the coefficient for as many times as the value for that case (e.g. age in years). But for a categorical variable the odds are for each category compared to one standard category (selected to be the final category unless we specify otherwise, as shown below).

The categorical variable Qualification has three categories. As the coding showed, the odds for each category are always relative to one of the other categories. Here the comparator is the default last category, having a degree. So the odds of 0.08 for compulsory education mean that, all other things being equal, someone in this dataset without any post-compulsory education is only 8% as likely to be employed as someone with a degree. Someone with post-compulsory education but not a degree is only 19% as likely to be employed as someone with a degree.

To complete this section and this simple model, note that as with linear regression it is possible to create an equation for the full model. This can be used to create scores in that model (the probabilities of being employed, for example) for each case. The computation would be the constant (here 1.939) minus 2.52 times variable (or category) 1, minus 1.67 times variable 2, and so on.

---

When running a logistic regression, click the Save option on the right of the main box. This will create a new variable for your dataset, based on the model predictions. You can choose to save the precise probabilities of group membership for each case, or just which group that case is predicted to be in.

---

## REFINING THE MODEL

It is also possible to plot the likelihood of cases being predicted to be in one or other of the two outcome categories. This plot can sometimes tell us something useful about the model or how to improve it (Figure 19.1)

Follow the steps for logistic regression as above, and in the dialogue window click the Options button, tick Classification plots, and click Continue. Click OK.

```
LOGISTIC REGRESSION VARIABLES Employed
/METHOD=ENTER Qualification Age
/CONTRAST (Qualification)=Indicator
/CLASSPLOT
/CRITERIA=PIN(0.05) POUT(0.10) ITERATE(20) CUT(0.5).
```

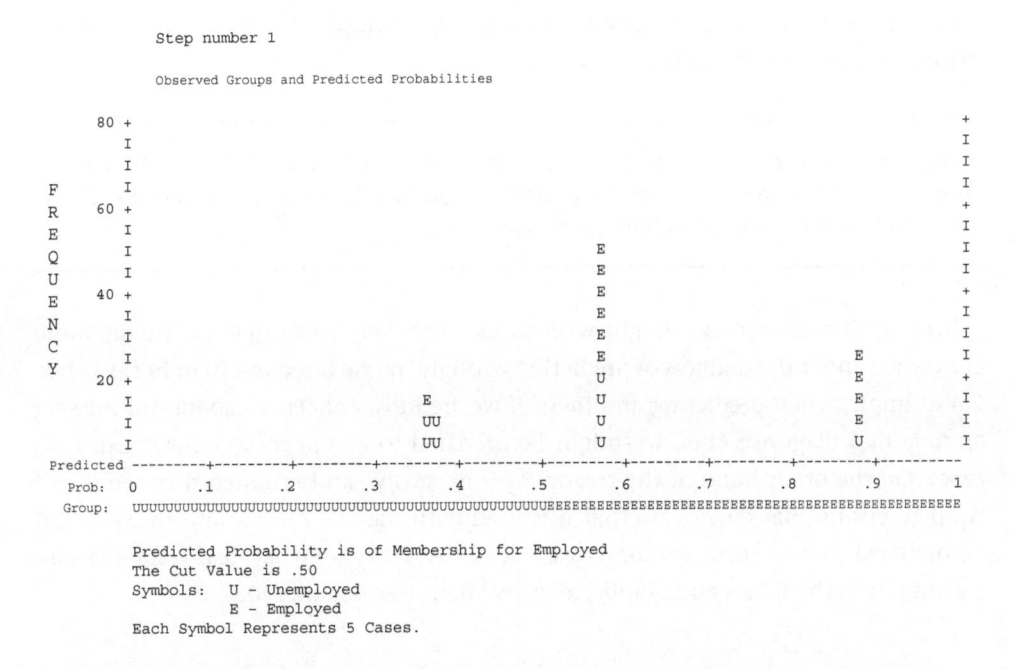

```
              Step number 1

              Observed Groups and Predicted Probabilities

        80 +                                                                    +
           I                                                                    I
           I                                                                    I
   F       I                                                                    I
   R    60 +                                                                    +
   E       I                                                                    I
   Q       I                        E                                           I
   U       I                        E                                           I
   E    40 +                        E                                           +
   N       I                        E                                           I
   C       I                        E                                           I
   Y       I                        E                              E            I
        20 +                        U                              E            +
           I              E         U                              E            I
           I              UU        U                              E            I
           I              UU        U                              U            I
   Predicted ---------+---------+---------+---------+---------+---------+---------+---------+---------+---------
   Prob:    0    .1   .2   .3   .4   .5   .6   .7   .8   .9   1
   Group:   UUUUUUUUUUUUUUUUUUUUUUUUUUUUUUUUUUUUUUUUUUUUUUUUUUEEEEEEEEEEEEEEEEEEEEEEEEEEEEEEEEEEEEEEEEEEEEEEEEEEE

              Predicted Probability is of Membership for Employed
              The Cut Value is .50
              Symbols:   U - Unemployed
                         E - Employed
              Each Symbol Represents 5 Cases.
```

**Figure 19.1** Classification plot for logistic regression model predicting employment

This slightly odd-format graph shows approximately where our 100 cases are, on a probability line from 0 to 1 of being employed, just using age and qualification as predictors. The x-axis is half labelled U and half E. All cases plotted over the U half are predicted to be unemployed, and all cases over the E half are predicted to be employed. The cut-off between U and E comes at 0.5 on the actual probability line (the cut point). Each E or U within the area of the plot is the actual outcome (not the predicted one), and each represents five cases. So we can see that some cases that are actually employed have been predicted to be unemployed in our model (where an E sits above the U half of the graph), and that even more unemployed have been predicted to be employed (where U sits above the E half).

What we are looking for in such a plot is how well the model discriminates between the two outcomes, in terms not only of the percentage allocated correctly, but also of how far apart they are. In Figure 19.1, lots of cases sit just one side or other of the 0.5 probability. This is not clear discrimination. It confirms that the 66% accurate model is not yet a good one. The cases at the more extreme end nearer 0.9 are better discriminated, but even these contain some that have been predicted wrongly (predicted U, but really E).

What we can sometimes do to improve the model using the plot is to envisage what happens if we move the cut point from 0.5 to elsewhere. For example, if we set up the model with 0.6 (60%) as the **cut-off**, we can see from the plot that at least 20 more unemployed cases (four Us) would now be correctly predicted as being unemployed. Unfortunately, this does not help because this would lead to at least 30 employed cases (six Es) now being incorrectly predicted as unemployed. So for this analysis, there is no gain in altering the cut point.

---

We could change the cut point when running the model by clicking the Options button on the main dialogue box (the same as when requesting a plot), and then typing a fraction (like 0.6) in the Classification cut-off box on the bottom right.

---

Even in this example we might wish to move the cut-off from 0.5 if we are more concerned about the dangers of predicting wrongly in one direction than in the other. For example, when predicting an illness, if we are most concerned about not missing anyone in a diagnosis then we might be prepared to accept more falsely identified cases. On the other hand, if the treatment is expensive and rationed then we might want to ensure that every case that it is used with has the illness, and so we might be prepared to miss some genuine cases. Such decisions are what 'real' data analysis, starting from the tables and graphs, is really about (see Chapter 8).

### Exercise 19.2

If we move the cut point to 0.3 in Figure 19.1, how many actually unemployed people would now be predicted to be employed?

---

Another way of looking at your data with logistic regression involves altering how the variables are introduced into the model – all at once (Enter), all at once and then removed one by one until their removal makes a difference to the percentage predicted correctly (Backward), or entered one at a time until a new variable makes no difference to the percentage predicted correctly (Forward). This is similar to what we tried with linear regression (Chapter 17). Both approaches should lead to a model that is as good as that for Enter, but which uses fewer variables (a more parsimonious model). Also, knowing which variables are unrelated to the outcome can lead to some very useful findings.

---

For example, in the SPSS dialogue box for logistic regression there is a dropdown menu towards the bottom labelled Method which is set to Enter. This means that all variables in the model will simply be entered at the same time. If you click the arrow by Enter, there are usually six choices – three labelled Forward and three Backward. In the Forward versions one variable is added to the model at a time until adding more variables produces no improvement in the model. In the Backward versions all variables are used (as with Enter) but then one variable at a time is removed from the model until the model gets noticeably worse. Each Forward and Backward choice also has a choice of how the system decides whether to add or remove a variable, such as conditionally, using the likelihood ratio, or the Wald statistic. Most of these involve significance testing and are perhaps best avoided.

---

Although it may be more effort, you can add variables yourself one at a time to see their effect size in the model. Does the percentage predicted correctly change, with the addition of that variable? If not, this suggests that the prediction does not need that variable. Try another variable. Doing it this way means you can understand more of what you are doing, and can use your judgement or a theoretical basis for any decision, rather than relying on any invisible 'magic'. Also, with syntax there is not that much effort involved anyway. Try it. It is important that your models are as simple as possible, and so have the fewest predictors.

Another possibility to improve the model is to examine interactions between variables (as with linear regression). It seems from the analysis above that age is not relevant to employment, once qualification is known. But the real relationship could be more complex. The average level of qualification may change with age, so it may be that age matters for some levels of qualification and not others (and vice versa). This is easy to assess (an interaction can be denoted by 'multiplying' two variables, like Age × Income).

---

Get the normal logistic regression syntax, and in the /METHOD line add the two variables you want to look at interaction for, and join them with a '*' – so here it could be Qualification*Age. If we also use the forward stepwise method for illustration:

```
LOGISTIC REGRESSION VARIABLES Employed
/METHOD=FSTEP (COND) Qualification Age Qualification*Age
/CONTRAST (Qualification)=Indicator
```

*(Continued)*

/CLASSPLOT

/CRITERIA=PIN(0.05) POUT(0.10) ITERATE(20) CUT(0.5).

Ignoring the clutter, and Block 0 which is the same as in the first example, the output with an interaction and forward entry of variables would look like this:

Classification Table[a]

|  |  |  | Predicted |  | |
|---|---|---|---|---|---|
|  | Observed |  | Employed | | Percentage Correct |
|  |  |  | Unemployed | Employed | |
| Step 1 | Employed | Unemployed | 16 | 25 | 39.0 |
|  |  | Employed | 9 | 50 | 84.7 |
|  | Overall Percentage |  |  |  | 66.0 |

a. The cut value is .500

Variables in the Equation

|  |  | B | S.E. | Wald | df | Sig. | Exp(B) |
|---|---|---|---|---|---|---|---|
| Step 1[a] | Qualification |  |  | 11.482 | 2 | .003 |  |
|  | Qualification(1) | −2.521 | .745 | 11.463 | 1 | .001 | .080 |
|  | Qualification(2) | −1.670 | .679 | 6.049 | 1 | .014 | .188 |
|  | Constant | 1.946 | .617 | 9.940 | 1 | .002 | 7.000 |

a. Variable(s) entered on step 1: Qualification.

Variables not in the Equation

|  |  |  | Score | df | Sig. |
|---|---|---|---|---|---|
| Step 1 | Variables | Age | .000 | 1 | .992 |
|  |  | Qualification * Age | .575 | 2 | .750 |
|  |  | Qualification(1) by Age | .194 | 1 | .660 |
|  |  | Qualification(2) by Age | .381 | 1 | .537 |
|  | Overall Statistics |  | 1.399 | 3 | .706 |

The analysis stops after entering only one variable. It shows a model with the same classification table as our original version (predicting 66% of cases correctly). This is because the simplest version of the model does not include age or the Age × Qualification interaction. These do not appear to improve the model at all, and so can be omitted. They are listed in the 'variables not in the equation' box. Age has been omitted. And the Age × Qualification interaction is also irrelevant. Only the simple level of qualification matters. This kind of simplification is important for reporting results (making models easier to explain), and if this were a real dataset the finding that age is not a factor could be at least as important as the rather duller finding that qualifications are linked to employment chances.

## TWO-STAGE LOGISTIC REGRESSION

Another way of investigating possible improvements in the model is to enter the predictors in separate stages or blocks (again as with linear regression). This is different from using Forward or Backward to determine how and whether predictors should be included in any one block (we can still use Enter, Forward or Backward in each block). But each block is considered as a useful separate step in the analysis.

For example, I have quite often entered predictor variables into a model in biographical order (where the cases are individuals). Block 1 might be everything we know about that person at age 6, Block 2 everything we know at age 11, Block 3 everything we know about them at age 18, and so on. In this way, we can get a sense of how much difference the later events make to the trajectory that the individual was already on from an earlier age.

The example below focuses on a two-stage model to help explain the idea as simply as possible. But a model like this can have many blocks or stages. For illustration, let us imagine that there is another variable in our made-up dataset that predicts employment. This new variable is a simplified one representing the individual's parental occupational class. We will use three categories for this illustration – the parent was never employed (or was in an otherwise unknown occupational group), or in a routine occupation, or a professional/supervisory one. They are coded in that order, with the last being the default category for comparison.

The idea underlying this example analysis is that the age of the respondent and the occupational class of their parents will pre-date the level of education attained by the respondent, in the sense that their age is set when they were born, and so on. So Age and Parentclass are entered as predictor variables in Block 1, and then the variable representing level of education is entered as a predictor in Block 2. So the order of the blocks is approximately biographical. In each block you can still change the method of entering the variables listed for that block, from Enter to Forward or Backward. And these can even differ for each block.

Any Block 2 variables can only explain variation left unexplained by Block 1 (and Block 3 only variation left after Blocks 1 and 2, and so on). This means that we can see the relevance of level of education, net of the relevance of age and parental class.

Enter the variables for the first step into the big blank area of the dialogue box as usual – here we use Age and Parentclass. Above this area it says Block 1 of 1. Click on the arrow to the left of the area, and the title is now Block 2 of 2. Move your desired variables for the second step into the newly blank area, and so on for as many steps as you want. Here we use level of education in Block 2. In each block you can still change the method of using those variables listed for that block, from Enter to Forward or whatever.

```
LOGISTIC REGRESSION VARIABLES Employed
/METHOD=ENTER Age Parentclass
/METHOD=ENTER Qualification
/CONTRAST (Parentclass)=Indicator
/CONTRAST (Qualification)=Indicator
/CRITERIA=PIN(.05) POUT(.10) ITERATE(20) CUT(.5).
```

The report in terms of variable coding and Block 0 would then be exactly the same as in the simpler model (still 59% correctly predicted at the outset). The simplified results for Blocks 1 and 2 are as follows. Table 19.4 for Block 1 shows that Age and Parentclass together produce predictions for current employment that are 70% accurate. This is a noticeable improvement on the 59% in Block 0.

**Table 19.4** Model from Block 1 of logistic regression predicting employment

|  | Predicted unemployed | Predicted employed | Percentage correct |
| --- | --- | --- | --- |
| Observed unemployed | 21 | 20 | 51.2 |
| Observed employed | 10 | 49 | 83.1 |
| Overall correct |  |  | 70.0 |

The corresponding coefficients in Table 19.5 show that Age now plays a small role in that prediction. Cases are 0.5% more likely to be employed (their odds are multiplied by 1.005) for every year of their age. However, the key potential determinant of employment, in this imaginary dataset, is parental occupational class. All other things being equal, those with unemployed parents have less than 10% odds of being employed (0.094), compared to those with parents with professional occupations. All other things being equal, those whose parents have a routine occupation have less than 30% odds of being employed (0.293), compared to those with parents with professional occupations.

**Table 19.5** Coefficients for Block 1 of logistic regression model predicting employment

|  | B | Exp(B) |
|---|---|---|
| Age | 0.005 | 1.005 |
| Parent unemployed/not known | −2.368 | 0.094 |
| Parent routine occupation | −1.228 | 0.293 |
| Parent professional/supervisory | − | − |

Table 19.6 is for Block 2 of our two-step model, with the addition of the individual's level of education. This improves the percentage of cases predicted correctly by 4 percentage points (to 74%), compared to Block 1. This increase is less than in the previous model, when we added level of education in the first and only block. There it added 6 percentage points. This is one indication of the potential importance of stepped models like this. Sometimes, later variables act as a proxy or substitute for earlier ones, and only by putting them in the model in a coherent, or appropriately theoretical, order can we see this. Here level of education would presumably act as a proxy for age and parent class if entered at the same time, and soak up some of the variation that the latter really explain. We know that age becomes apparently irrelevant when entered with qualification. This new model suggests that age does have a minor role, as well as qualification. Separated out in this manner, all three variables seem to contribute something to the final result.

**Table 19.6** Model from Block 2 of logistic regression predicting employment

|  | Predicted unemployed | Predicted employed | Percentage correct |
|---|---|---|---|
| Observed unemployed | 27 | 14 | 65.9 |
| Observed employed | 12 | 47 | 79.7 |
| Overall correct |  |  | 74.0 |

Table 19.7 shows the odds for all three variables in the final step. The odds for age and parent class are similar to those in Table 19.5 but have been modified a bit by the inclusion of level of education. This is normal. The odds for all remaining variables will tend to change slightly as new variables are entered or removed. As in the previous example, and as expected, a lower level of education is associated with considerably lower odds of being in employment. Here the odds are net of age and parental occupation.

**Table 19.7** Coefficients for Block 2 of logistic regression model predicting employment

|  | B | Exp(B) |
|---|---|---|
| Age | 0.014 | 1.014 |
| Parent unemployed/not known | −2.431 | 0.088 |
| Parent routine occupation | −1.178 | 0.308 |

(Continued)

**Table 19.7** (Continued)

|  | B | Exp(B) |
|---|---|---|
| Parent professional/supervisory | – | – |
| Compulsory education | –2.377 | 0.003 |
| Post-compulsory education | –1.990 | 0.008 |
| Degree or higher | – | – |

The examples so far have been realistic but imaginary. We cannot draw any useful implications from the results because the data has been made up.

## CHANGING THE REFERENCE CATEGORY

In any logistic regression it is possible to change the reference category for categorical variables (the one missing a value in the table). The default is the last category. Changing this does not alter the substantive result or what it means, but it can make things easier to explain and understand. For example, imagine that the parent variable had categories of employed or unemployed, but with a third category of not known. Then the odds for employed and unemployed would be reported relative to the cases with an unknown occupational status. This might look odd. If you prefer to compare employed and unemployed more directly then you could change the reference category to one of these. You could do this either by recoding your data (in SPSS, for example), or by picking a different reference category when you specify which variables are categorical.

---

Select the appropriate variable once you have put it in the Categorical Covariates box, and you will see below this the options to tick in order to change the reference category. The simplest step is to make the first category the reference, instead of the last category.

---

## THE POSSIBLE DETERMINANTS OF ENTERING HIGHER EDUCATION: A REAL EXAMPLE

The final example in this chapter comes from a real research study. This was an attempt to identify the possible determinants of young people entering higher education (university, henceforth HE) or not (Gorard, 2018). The dataset was a linked version of the National Pupil Database and the Higher Education Statistics Agency database on university entrants for England. The number of young people with sufficiently complete records is around 580,000 for one annual cohort, of which around 280,000 are in a position to apply to enter HE. The binary outcome variable is whether a young person was accepted for university or not (around one-third were accepted in that year, at or near the traditional age of 18). The predictor variables included prior attainment from the age of 13 onwards, and a range of individual background variables such as ethnicity, sex, poverty, first language, and whether the young person had been living in state care.

The purpose of this model was to help judge whether the clear stratification of the student body in England, in terms of poverty, ethnicity and so on, was due chiefly to prior differences in attainment, or whether there is additional unfairness facing disadvantaged students on applying to university. The answer is important for where the state might wish to prioritise policy and spending. If students are accepted to university in proportion to their qualifications then anyone wishing to widen access has to reduce the role of qualifications in deciding on places, and/or emphasise work on the destratification of prior qualifications at school.

On the other hand, if students are not accepted to university roughly in proportion to their qualifications, then this could be evidence of unfairness in admissions policies and practice as well. This unfairness could be addressed by legislation and enforcement, as well as, or instead of, work at the school level. Because this was the purpose of the analysis, the predictors are not entered in biographical order (as above). Instead the qualification data is entered first, so that the potential link to student background could be examined net of qualifications.

Around 67% of young people did not attend university in England at the traditional age – whether because they were not interested, did not apply, or were not accepted. So the base model (Block 0) is 66.7% accurate already (Table 19.8). Accuracy rises by 9.6 percentage points when prior qualification is added to the model in the first step. And student background then only adds another 1.8 percentage points. So, students appear to be being selected largely on the basis of prior qualifications, although it is likely that these qualifications are themselves predicated to some extent on student background.

**Table 19.8** Percentage predicted correctly for HE entry, step by step

| Step | Enter HE |
| --- | --- |
| Base | 66.7 |
| Step 1: KS4 and KS5 attainment | 76.3 |
| Step 2: Student background (school) | 78.1 |

The odds of a young person entering university are positively related to their attainment (Table 19.9). For example, their chances increase by 5% (are multiplied by 1.05) for every A or A* grade they achieve at Key Stage 4 (aged 16). The small number who had been living in care are underrepresented to a greater extent. Lower-income students (eligible for free school meals) appear to participate in almost direct proportion to their earlier qualifications.

On these figures, ethnic minority students (all categories) and those with English as an additional language are considerably overrepresented in HE compared to white and English first-language young people. For example, once their qualifications and other factors are taken into account, Black students are five times as likely as white students to enter a university of some kind.

These are remarkable figures, and should be treated with caution. This is only part of the analysis, presented as an example for tutorial purposes. Black students, like all potentially disadvantaged students, are still less somewhat likely to attend the more prestigious institutions and courses, even with equivalent qualifications. They are slightly more likely to drop out, and less likely to leave with a higher degree classification (as discussed in Gorard, 2018). The battle against stratification has many phases.

Combined with other findings, the figures suggest that attempts were already being made to make the university application process fairer, and even to compensate or adjust for student background differences in prior attainment. The stratification problem for HE more clearly lies in the stratification of prior attainment and less in unfairness in the selection of students later on. For example, and more problematically, Black students will probably still be underrepresented in HE because of differences in prior attainment, even if those who present themselves as applicants are now more likely to be accepted. This suggests that the focus for improvement should be in schools and wider society, and not on the admissions process more generally.

Table 19.9  Coefficients for variables in final step of model predicting HE entry

| Variable | Step 2 odds |
| --- | --- |
| Key Stage 4 capped points score | 1.01 |
| Number of A and A* grades at A level | 1.05 |
| Number of passes A* to C at A level | 1.06 |
| Key Stage 5 total points | 1.004 |
| In care (v. not) | 0.80 |
| Not FSM-eligible (v. eligible) | 1.02 |
| English as an additional language (v. not) | 1.99 |
| Special educational need (v. not) | 0.72 |
| Any 'other' ethnicity (v. white) | 3.42 |
| Asian (v. white) | 3.83 |
| Black (v. white) | 5.02 |
| Chinese (v. white) | 3.53 |
| Mixed (v. white) | 2.04 |

The point of this example is to illustrate logistic regression being used in a relatively simple way, with existing valuable datasets, to address a real problem, and with results that have possibly important implications for current policies, such as contextualised admissions to university.

## Exercise 19.3

In Table 19.9 the odds of entering university based on prior Key Stage 4 attainment score are 1.01, and based on not being eligible for free school meals they are 1.02. This means that it is more important not to be poor (not FSM-eligible) than to do well in exams. Discuss.

## CONCLUSION

This chapter has focused on simple binary logistic regression (only two outcomes). There are, in addition, more complex approaches for multiple category outcome variables, and for ordinal variables (Allison, 1984; Mare and Winship, 1985). And there are more recent developments for both real number and categorical variables combined, such as regression trees.

In my experience, the percentage of cases predicted correctly does not rise much above the base value whenever a binary logistic regression starts from a base value very far from 50%. If, for example, the base is 90% it is very hard to create a model with real-life variables that can do better than this – where a pure guess would already be 90% accurate. And multiple-outcome (multinomial) logistic regression seldom has a near-even spread of cases between all outcome categories at the outset. It is usually safer to run a series of binary logistic regressions when you want to examine more than two categories of outcome. See, for example, the models in Gorard (2017). Watch out for this problem in the literature, where there is a deplorable habit of not reporting the percentage of cases predicted correctly (or anything similar, such as a pseudo-$R^2$). Without this key fact we have no idea how much the coefficients/odds really explain.

Whatever kind of logistic regression you want to conduct, there are things that can be done if your cases are very unbalanced at the outset (e.g. if you are predicting who will fall seriously ill in the next month, there is considerable imbalance, because most people will not fall ill). In a binary analysis you could use the smaller group as one group, and create a randomly selected subgroup of the larger group, to be the same size. Your base is then 50:50. You could even do this many times, and average the ensuing results. This is not an ideal solution, and should only be done when you have a large number of cases even in the smaller category. But it is more fruitful than starting with a base proportion of 98:2, for example.

Whatever kind of logistic regression you conduct or are reading, remember that the percentage of cases predicted correctly should be compared to the base figure, and not to zero or 100%. A model that starts with a base of 50% and predicts 75% of cases correctly is a potentially strong one. A model that starts from a base of 72% and predicts 75% of cases correctly is probably a trivial one. If a report does not state this information clearly, as most still do not, then do not trust it.

## Notes on selected exercises

### Exercise 19.1

This cell and the third cell (first cell, second row) should both contain 0 cases. All cases would be predicted correctly.

### Exercise 19.2

There would be four extra U symbols over the E part of the axis, so there would be 20 new misclassifications. There would be five extra correct predictions of unemployment, but overall moving the cut point to 0.3 makes the model worse.

### Exercise 19.3

Not being poor may well be more important than qualifications in general, but that is not a correct interpretation of the odds in Table 19.9, for entry to university at least. The FSM variable is categorical and binary. Someone who was not classified as poor would have their chances multiplied by 1.02 (increase of 2%). That is all. The attainment variable is a real number. An individual has their chances multiplied by 1.01 (increased by 1%) for every Key Stage 4 point that they score. In this table, attainment is far more important than poverty as a predictor.

## Further exercises to try

Find a suitable real-life dataset you are interested in (or use one provided on the website). Select or create a binary categorical variable as the outcome, and use any other variables as predictors in a simple logistic regression analysis. Make sure you specify which of the predictors are categorical. What is the difference between the percentage of cases predicted correctly in Blocks 0 and 1? Try it again with two categorical predictors in interaction. Has this made any difference?

## Suggestions for further reading

A short primer on logistic regression:
Menard, S. (1995) *Applied Logistic Regression Analysis*. London: Sage.

A classic study that used logistic regression in an original way (ignore any reference to significance tests):
Gambetta, D. (1987) *Were They Pushed or Did They Jump? Individual Decision Mechanisms in Education*. Cambridge: Cambridge University Press.

A well-received study in the sociology of education, based largely on logistic regression:
Gorard, S. and Rees, G. (2002) *Creating a Learning Society*. Bristol: Policy Press.

# 20
# DATA REDUCTION TECHNIQUES

## SUMMARY

This brief final substantive chapter on analysis introduces the idea of data reduction, with a focus on factor analysis for data reduction or simplification, and reliability coefficients as used in developing questionnaire instruments. The techniques in this chapter could be used to address research questions such as:

- Is there good evidence of a more generic concept underlying individuals' responses to several questionnaire items?
- Which of several questionnaire items is the strongest in picking up a particular concept?
- How 'reliable' is a set of questionnaire items?

## WHAT IS FACTOR ANALYSIS?

Factor analysis is a name given to a range of related techniques used to portray possible hidden measures underlying sets of observable questionnaire responses. These are sometimes called **latent** variables. Using factor analysis can help to simplify your dataset, by reducing the number of variables to consider in any subsequent analyses (Gorsuch, 1972), while attempting to retain the majority of the important variation in the original dataset (Marradi, 1981).

### Assumptions underlying factor analysis

Factor analysis is like a multiple correlation, and has some of the same basic assumptions – such as that all of the variables involved must be real numbers. In addition, but in common with many multivariate analyses, there must be substantially more cases than initial variables in the model. Some resources suggest a minimum of at least 10 times as many cases as variables, but usually more. There is no special threshold. It should just be clearly more. So a factor analysis of 10 variables based on 1,000 cases is OK; a factor analysis of 37 variables based on 100 cases is not.

## A WORKED EXAMPLE OF FACTOR ANALYSIS

This technique is probably best understood via an example. This section presents some simplified findings from a real-life study of career choice and teacher supply in England (Gorard et al., 2020). A total of 4,469 home undergraduates in 53 universities in a non-probability sample responded to a questionnaire which included 20 items about the issues that the students may or may not consider when choosing a career. Students were asked to rate (out of 10) the importance of issues such as pay and conditions, job satisfaction, and intrinsic motivators. Here we will look at how interlinked these disparate items appear to be.

There are different versions of factor analysis, based partly on precisely how the factors are extracted (Maxwell, 1977). The example uses the principal components method, but because the results are 'rotated' to find the best solution, the method used initially makes little difference by the end (Comrey, 1973). The example requests varimax orthogonal rotations, making the factors as independent of each other as possible (Child, 1970).

---

From the "Analyze" menu select "Dimension Reduction", and then "Factor". Move the full set of real number variables you want to use to the Variables box. Click the Rotation button and select Varimax. Click OK.

FACTOR

/VARIABLES Pay Security Autonomy Prospects Opportunity Responsibility

Academicknowledge Chancegive Interest Colleagues Temperament Share Workload Familytrad Status

Length Stimulation Incentive Bonus Internship

/MISSING MEANSUB

/ANALYSIS Pay Security Autonomy Prospects Opportunity Responsibility

Academicknowledge Chancegive Interest Colleagues Temperament Share Workload Familytrad Status

Length Stimulation Incentive Bonus Internship

/PRINT ROTATION

```
/FORMAT SORT BLANK(.5)
/PLOT EIGEN
/CRITERIA MINEIGEN(1) ITERATE(25)
/EXTRACTION PC
/CRITERIA ITERATE(25)
/ROTATION VARIMAX
/METHOD=CORRELATION.
```

The first part of the output looks like this 'scree' plot in Figure 20.1. We will not worry much about technical terms like 'eigenvalue' here. This graph is not essential. What it shows is the amount of variation in the responses (y-axis) that can be explained by each underlying factor (consisting of contributions from several variables). Successive factors become less and less useful in explaining the common variance. Factors are extracted until the residual correlation is too close to zero, but generally as many factors are extracted as there are initial variables (here 20), in order to explain all of the variance possible.

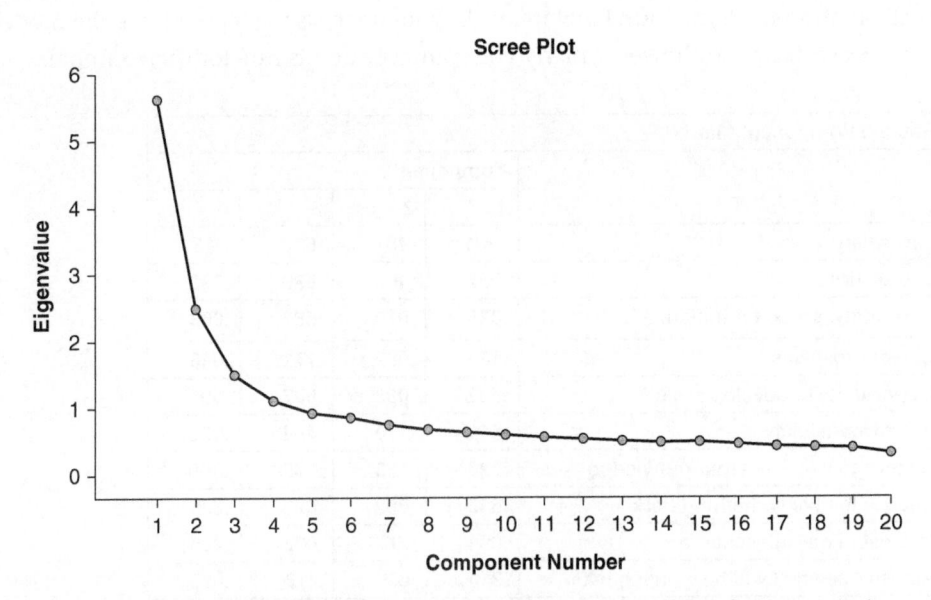

**Figure 20.1** Amount of variance covered by each factor in a simple factor analysis

However, unless the measurement of the variables is wholly reliable, at least some of the variance is due to error. To attempt to explain all of the variance, including that due to error, is unparsimonious, and the number of factors used should be kept to a reasonable minimum (Cureton and D'Agostino, 1983). After a few factors the new variation explained by each added factor tends to settle down at a low base. There is no point in using 20 factors to explain 20 items, because we would not have simplified the situation at all. The cut-off for adding to the model but keeping the number of factors low in this example would appear to be somewhere between 4 and 6 factors, where the graph really begins to flatten out.

Total Variance Explained

| Component | Rotation Sums of Squared Loadings |  |  |
|---|---|---|---|
|  | Total | % of Variance | Cumulative % |
| 1 | 3.269 | 16.345 | 16.345 |
| 2 | 3.038 | 15.192 | 31.537 |
| 3 | 2.391 | 11.957 | 43.494 |
| 4 | 2.072 | 10.359 | 53.853 |

Extraction Method: Principal Component Analysis.

This second part of the output shows that the model has stopped at four factors. The percentage of variance explained by each factor declines from over 16% to 10%, and the overall model retains only 54% of the variation in the 20 items. This is not unusual. There is a trade-off between a simple model (parsimony) and retaining the most variation from the individual items. In your analysis you could force the model to use 3 or 5 factors or however many you wanted, but 4 is fine for this example.

Rotated Component Matrix[a]

|  | Component |  |  |  |
|---|---|---|---|---|
|  | 1 | 2 | 3 | 4 |
| Pay, salary | -.311 | .408 | .536 | .113 |
| Job security | .037 | .160 | .539 | .339 |
| Autonomy, scope for initiative | .375 | .070 | .587 | .063 |
| Career prospects | .189 | .089 | .771 | .046 |
| Opportunity to develop skills | .532 | .033 | .537 | .001 |
| Job responsibility | .519 | .170 | .464 | .032 |
| Chance to use academic knowledge | .723 | .163 | .139 | .000 |
| Chance to give something back | .652 | .002 | .028 | .306 |
| Interest in my subject area | .684 | -.029 | .032 | .205 |
| Kinds of people I will be working with | .310 | .030 | .112 | .695 |
| Job that suits my temperament | .282 | .043 | .105 | .673 |
| Chance to share my knowledge | .696 | .157 | .083 | .309 |
| The workload required | .109 | .346 | .075 | .624 |
| Family tradition | .209 | .677 | -.123 | .048 |
| Status, public perception of the job | .095 | .643 | .119 | .130 |
| Length of working day, holidays | -.096 | .510 | .071 | .553 |

| | | | | |
|---|---|---|---|---|
| Intellectual stimulation | .516 | .154 | .336 | .046 |
| A financial incentive to train | .058 | .608 | .345 | .162 |
| An introductory bonus when starting job | -.020 | .780 | .130 | .167 |
| Opportunity for internship | .148 | .703 | .172 | -.009 |

The third part of the full output to look at is the final factor matrix above. This lists the 20 items from the survey (rows), and the four proposed components that underlie them and explain 54% of their variation (the columns). The figures in the table are the correlations between the responses to each survey question, and the proposed components or factors. You will see these correlations referred to as 'loadings', which is correct, but imagine them as correlations between −1 and +1, so the square of a loading can be seen as the amount of variance common to both. The first item, pay, is negatively related to the first factor, and positively related to the next three. Its strongest link is to factor 3.

The task now is to look at the pattern of links and try to characterise each factor (and presumably give each of these latent variables a name). With 20 items and 4 factors this is quite a task. It can be made easier by eliminating the smaller numbers (small in absolute terms, but keeping the large negative values). This can be specified in your software when running the model, or by hand afterwards.

If we ignore any value less than 0.5 in absolute value, we get Table 20.1. The table not only has no small numbers, but also groups together the items with high loadings for each factor. This makes the position much clearer to the reader (Chapter 21). Pay is now linked only to factor 3, by its highest loading (0.536). The first group (factor 1) seems to be about intrinsic interest in the job and its values. The second is more about incentives, but some more thought would be needed to sort this out further. The third is about pay and promotion, and the fourth is about job conditions. Although this was never intended to be an analysis used in the real study, this one works out quite neatly. We could say that there are four main factors underlying career choice (and that different types of respondents value them differentially).

**Table 20.1** A four-factor model of career choice influences

| | Factor 1 | Factor 2 | Factor 3 | Factor 4 |
|---|---|---|---|---|
| Chance to use academic knowledge | 0.723 | | | |
| Chance to share my knowledge | 0.696 | | | |
| Interest in my subject area | 0.684 | | | |
| Chance to give something back | 0.652 | | | |
| Job responsibility | 0.519 | | | |
| Intellectual stimulation | 0.516 | | | |

*(Continued)*

**Table 20.1** (Continued)

|  | Factor 1 | Factor 2 | Factor 3 | Factor 4 |
|---|---|---|---|---|
| An introductory bonus when starting job |  | 0.780 |  |  |
| Opportunity for internship |  | 0.703 |  |  |
| Family tradition |  | 0.677 |  |  |
| Status, public perception of the job |  | 0.643 |  |  |
| A financial incentive to train |  | 0.608 |  |  |
| Career prospects |  |  | 0.771 |  |
| Autonomy, scope for initiative |  |  | 0.587 |  |
| Job security |  |  | 0.539 |  |
| Opportunity to develop skills | 0.532 |  | 0.537 |  |
| Pay, salary |  |  | 0.536 |  |
| Kinds of people I will be working with |  |  |  | 0.695 |
| Job that suits my temperament |  |  |  | 0.673 |
| The workload required |  |  |  | 0.624 |
| Length of working day, holidays |  | 0.510 |  | 0.553 |

To get a sense of what factor analysis is doing here you could try just running the bivariate correlations between each of the 20 original items (as in Chapter 15). Again, ignore the small absolute correlations (perhaps those below 0.5). Then link the items together in groups with the items to which they are most strongly correlated. This yields strings of items that are similar to the 'factors' above. For example, all of the job satisfaction and altruistic (a chance to …) items form a group when analysed like this, and workload and length of working day form the basis of another. This simple correlational approach tends to lead to slightly more groups than factor analysis. But both approaches can produce quite different models anyway, if you change just one or two elements (adding or removing an item, forcing the model to have a fixed number of groups, and so on).

As with other models (see Chapters 17 and 19), you can save the new scores for each respondent, representing their scores for each factor, thereby converting your 20 measures into just 4. These scores are defined as the sum of the case score on each variable, multiplied by the loading for that variable on the factor, for all relevant variables (Jackson and Borgatta, 1981). You can then do further analyses with these latent factor scores, such as looking at correlations or differences between groups of respondents, just as you would with the original surface scores.

Another more radical approach, if you want to simplify your items into fewer factors, is to use only one item from each factor in your next questionnaire version. Look at factor 3. Only career prospects has a substantial loading on this factor (an $R^2$ of 0.5 or more). You could be ambitious, just using this item to represent the factor, and ditching the other items in this factor group.

## Concluding warning

However, the project from which this data comes did not use any of the ideas in this chapter. This is partly because factor analysis wastes so much potentially useful information. Just because an item or a part of the variation in an item does not correlate well with other items in a questionnaire, this is no reason to discard it. Rather it could be a reason to retain it. It may be asking something new and different. Of course it may not be relevant (valid), but this is very different from not being reliable (in this restricted sense of having responses linked to other items). See Chapter 14. I have only occasionally found a reason to use a full factor analysis in real research.

# RELIABILITY

Chapter 14 discussed the idea of reliability, and the use of repeated versions of the same/similar questions, when constructing a psychological attitude scale or similar. For example, an organisation might give a job satisfaction survey to its employees with several items about job satisfaction, rather than just one. This approach was not recommended in Chapter 14, on logical and data quality grounds. Such scales routinely discuss their version of 'reliability', by which they mean the extent to which different items asking what is in effect the same thing yield consistent responses from each individual. And this reliability is usually expressed in terms of **Cronbach's alpha**.

Cronbach's alpha is a measure of internal consistency, or how closely related a set of questionnaire items are, assessed in terms of the pattern of responses for each respondent (Cronbach, 1951). The coefficient (alpha) ranges from an unlikely 0 (no consistency) to 1 (the pattern of responses to all items is identical for each respondent). In some ways it is similar to the results from a correlational or factor analysis. If the items correlate more highly then alpha will be higher, and vice versa. If there are more items, and they correlate among themselves, then alpha will be higher still.

A high reliability is generally taken to mean that most/all of the items are on task, in the sense of measuring what is intended to be measured. Of course it does not mean anything of the sort. If all of the questions were about self-esteem this might yield a high alpha, even if the questions were meant to be about enjoyment, for example. Validity is the term used to describe whether the measure is any good at measuring what it is supposed to. Reliability is the term used here to describe whether the items in the measure are mostly related to each other, and about the same thing. These two aspects (validity and reliability) may be linked, but they are not necessarily so. You could also run a factor analysis (see above) and see if more than one underlying dimension turns up. If it does, this suggests that your items are actually about two or more things.

## 252 | HOW TO MAKE SENSE OF STATISTICS

Methods resources offer different values which we would want alpha to be, just as with correlations (Bland and Altman, 1997). A value of less than 0.5 may be considered poor, above 0.75 OK, and above 0.9 excellent, but these values are entirely arbitrary.

Using the same 20 items as in the factor analysis section above, about the issues reported as influencing career choice among undergraduates, we could also conduct a reliability analysis. This would help us look at the extent to which the items were asking the same thing in different ways.

---

From the "Analyze" menu select "Scale" and then "Reliability Analysis". Move the group of real number variables you want to consider to Items.

```
RELIABILITY
/VARIABLES=Pay Security Autonomy Prospects Opportunity Responsibility
Academicknowledge Chancegive Interest Colleagues Temperament Share Workload
Familytrad Status Length Stimulation Incentive Bonus Internship
/SCALE('ALL VARIABLES') ALL
/MODEL=ALPHA.
```

---

The output confirms that there are 4,469 cases and 20 items. The resulting alpha coefficient is 0.86, suggesting good reliability, or a high degree of linkage between the items (you could treat it a bit like a multiple correlation coefficient). But the items in this questionnaire were not devised to be similar. Rather they were meant to represent different issues that respondents might consider when choosing a job – as different as pay and academic interest are. These differences are clear in the patterns of responses for different items (see the analysis in Gorard et al., 2020), and the factor analysis (above) presented the responses as suggesting at least four distinct ('orthogonal') factors. So what has happened?

| Reliability |   |   |   |
|---|---|---|---|
| Scale: ALL VARIABLES |   |   |   |
| Case Processing Summary |   |   |   |
|   |   | N | % |
| Cases | Valid | 4469 | 100.0 |
|   | Excluded[a] | 0 | .0 |
|   | Total | 4469 | 100.0 |
| a. Listwise deletion based on all variables in the procedure. |||| 

| Reliability Statistics |   |
|---|---|
| Cronbach's Alpha | N of Items |
| .861 | 20 |

First, there are 20 items and alpha is sensitive to the number of items, so that it is bigger if there are more items (though I have never fully understood why this should be). Second, the factor analysis shows that there is some interconnectedness, and several items have a general theme. However, it is also possible that this shows never to read too much into a reliability analysis! I would usually prefer using correlations, regression or even factor analysis to look at the interrelatedness of items.

## CONCLUSION

Factor and reliability analyses are commonly encountered together when developing and testing questionnaires for latent variables like attitude scores. There are some slight differences in their use. For example, when developing an instrument from scratch you are more likely to let the data, rather than theory, decide how many factors there should be. But the principles outlined in this chapter also apply to considering an existing instrument. The existence of this chapter should not be seen as encouragement to undertake such analyses (see Chapter 10). This brief chapter is really here to give readers an outline idea of what other authors may be talking about.

There is a range of other possible approaches to reducing a set of variables by one or more dimensions. For example, cluster analysis groups categories of things together in terms of their characteristics (Everitt, 1980). Multi-dimensional scaling (MDS) is a similar technique, not used much, but very satisfying when it works. Like factor analysis, it involves looking at patterns in responses over many variables or questionnaire items, and sorting them into two or more dimensions. Like all of these techniques, MDS does not always work. For a reasonably successful example, see Gorard with Taylor (2004).

### Suggestions for further reading

Two very brief books or chapters to help you if you want to pursue the idea of factor analysis:
Kim, J. and Mueller, C. (1978) *Introduction to Factor Analysis: What It Is and How to Do It*. London: Sage.

Marradi, A. (1981) Factor analysis as an aid in the formation and refinement of empirically useful concepts. In D. Jackson and E. Borgatta (eds), *Factor Analysis and Measurement*. London: Sage.

A relatively simple description of a few types of reliability:
McLeod, S. (2013) What is reliability? *Simply Psychology*. https://www.simplypsychology.org/reliability.html

# Part V

# Conclusion

# 21
# PRESENTING DATA FOR YOUR AUDIENCE

## SUMMARY

Having looked at many examples of analyses, and the sometimes bewildering output that can be generated by analytical software, this chapter offers additional advice on how best to present your numeric results. Examples and advice appear in just about every chapter of the book, and more are illustrated here in one place for convenience.

When we and others report our research, our purpose must be to explain it as simply as possible to gain the widest possible comprehending readership. This permits people to appreciate and use our research findings, leading to the widest possible opportunity for critique or replication, and so to improvements in our field of research. The act of converting our findings into a simple report format also helps us to understand our own work and its limitations better. This brief chapter suggests ideas about the clear presentation of numeric results, although many of the points made would also apply to reports not based on numbers.

## INTRODUCTION

### Exercise 21.1

Consider this computational problem. In a standard knockout competition, such as a singles tennis tournament, if there are four players then there will be three matches in total - two first-round matches and a final. If there are eight players there will be seven matches in total - four first-round matches, two second-round matches and a final. It seems that for any number ($n$) which is a power of 2, there will be $n - 1$ matches in total. But if $n$ is not a power of 2, then unmatched players will have a bye in the first round. How many matches in total would be needed for a tournament of 43 players? How could you prove your answer for any number of players?

This seems quite daunting. Before you start working this out with paper and pencil, consider the following. Each match (other than a bye) has two players, and only one player goes through. So each match eliminates one player. To have a winner we need to eliminate all but one player. Therefore, *however* many players there are, the tournament always needs $n - 1$ matches (so, 42 matches for 43 players). This is a very simple proof of the general solution, and most people can recognise it as such. But in practice even professional mathematicians can struggle to find the proof for themselves (Dawes, 2001). This is perhaps because the initial problem is phrased in such a complex way that many readers start looking for a complex answer. Once the problem is phrased more simply, the solution is obvious. My point here is that when dealing with difficult issues in our research we should try and represent them as simply as possible. This, in itself, may not be easy. It can take hard mental work. But once done, the simpler representation is easier to work with, easier to communicate, and easier to teach.

## Exercise 21.2

As another example, imagine being faced with the following realistic problem. Around 1% of people will get a specific disease. If someone is going to get the disease, then they have a 90% probability of obtaining a positive result from a diagnostic test. Those people who will not get the disease have only a 10% probability of obtaining a positive result from the diagnostic test. The test is therefore quite accurate (90% in either direction). If everyone in a region is tested, and a person you know has just obtained a positive result from the test, then what is the probability that they will get the disease?

Faced with problems such as these, many people are unable to calculate a valid estimate of the risk. This inability applies to relevant professionals such as physicians and counsellors, as well as researchers and statisticians (Gigerenzer, 2002). Yet such a calculation is fundamental to the assessment of risk in real-life situations. Many people might think the chances are about 90% (the accuracy of the test). These people have confused the conditional probability of someone getting the disease, given that they have a positive test result, with the conditional probability of getting a positive test result, given that someone will get the disease. As shown below, and as was shown in relation to *p*-values in Chapters 6 and 7, the two values are completely different.

Rather than guessing or doing complex calculations, a better approach is just to express the problem in an easier way. Consider the same problem expressed as frequencies rather than as probabilities.

Of 1,000 people chosen at random, on average 10 will get the disease. Of these 10 people, around 9 will obtain a positive result in the diagnostic test. Of the 990 tested who will not get the disease, around 99 will also obtain a positive test result. These figures are summarised in Table 21.1. They are the same figures as above: 10 is 1% of 1,000, 99 is 10% of 990, 990 is 1,000 minus 10, and so on. If all 1,000 people are tested, and someone you know is one of the 108 obtaining a positive result, what is the probability that they will get the disease?

Table 21.1  Probability of getting the disease, having tested positive

|  | Test positive | Test negative | Total |
| --- | --- | --- | --- |
| Will get disease | 9 | 1 | 10 |
| Will not get disease | 99 | 891 | 990 |
| Total | 108 | 892 | 1,000 |

This is a much easier question. Of the 1,000 people, 108 will test positive but only 9 of these will actually go on to develop the disease. This means that the probability of developing the disease after a positive test result is 9/108 or just over 8%. This would still be alarming, but 8% is very different from near 90%. Changing the description makes the solution far easier to see.

Exactly the same confusion occurs when researchers imagine (or pretend) that the probability of seeing a pattern in some data, assuming that there is no pattern really (the *p*-value from significance tests), is the same as the probability of there being no pattern in the data, given the data they obtained (what they really want to know). One figure can be big and the other small, and there is no way of telling without also knowing the underlying non-conditional probability (like the 1% in this example), and applying the same process as here.

The idea behind both examples so far is that expressing difficult ideas involving numbers carefully, simply and sometimes differently than 'usual' can revolutionise how easy they are to understand. We can apply this principle to our research reporting.

## GENERIC ISSUES OF PRESENTATION

The first and most important issue in presenting research results is clarity of writing. If a research report is easily readable then it becomes easier to judge the information within it (Chapter 8). This means using no long words unless absolutely necessary, no unfamiliar new terms, and no long sentences and paragraphs (the overuse and incorrect use of semicolons is often a symptom of this). The job of the writer is to work hard to help the reader to read their writing.

Research reports must also be complete in the sense that they must describe the research questions, research design, methods of data collection and analysis, clear results, limitations, and suggested implications. There is further discussion of this issue in Chapter 8.

Here is a real-life abstract to illustrate the importance of clarity. It appeared in a high-ranked journal, and is shown here only as an example of a wider pattern. The article has the same style as the abstract throughout.

This is a heterodidactic neuroscience, not materiality of language discussion. It is an article on intra-observation as research methodology, thus auditory, visual, cultural, syntactic, and paratactical body/mind/brain possibilities of words – Thinking ontology for post-qualitative methodology. Indirectly, it is about linguistic accountability of the event of/in education; the event of/in research – Thought in the act and/or bodily awareness in the moment: Nodal point waiting pentimento performances .... and Jasmine. It might therefore be about a whole new transdisciplinary and humble body/mind/brain, education and research field all together, or about being a bit wild in the brain, and/to transfer something across areas and learning. It is a schizoanalysis crafting of knowledge attempt becoming data wise in action. My goal and hope are to foster self-reflexive think-languages and practices in Early Childhood Education and Care (ECEC) Institutions and schools. It is a professionalizing method. (Reinertsen, 2015)

The abstract makes no sense to me. It conveys no idea of what the research is in terms of the issue addressed, research design, methods, results or implications. Perhaps it is not summarising an empirical article but a conceptual piece. There is no obvious way of knowing. This is the full abstract. It appears to be made deliberately hard to read, and responsibility for the difficulty lies with the author, not me as the reader. In any review of evidence I would just ignore it. This abstract is not that unusual, and the phenomenon of writers not caring to help their wider readership occurs across all social science.

## PRESENTATION OF NUMBERS

The purpose of reporting research is to enable the widest possible readership to understand what you have done and what you found out. Having looked at examples of simplification and of poor writing, the chapter turns to some advice on presenting numeric results so that the widest readership can understand them. This advice appears in no particular order, is not exhaustive, and may well not be the best imaginable. The key point is to care, and to think about presentation.

Perhaps the most obvious place to start with numbers is how long they are. Longer numbers are harder to read, just like longer words and sentences. We cannot control the scale of the numbers in our research, but we can present them succinctly. Whole numbers can be described in terms of hundreds, thousands or millions. Decimal fractions should appear with a sensible number of **significant figures**. In social science most measurements are not tiny decimal fractions, so this is really a question of using the minimum number of decimal places necessary (see the real-life examples later in this chapter, and others throughout the book).

If you have created a sample of 10,000 cases, and measured some characteristic with high quality (perhaps the number of people resident in each household), with very low non-response, then it makes sense to portray that accuracy in your results. You might report that the average number of residents per household was 2.74 (accurate to two decimal places, or five-thousandths of a unit). This is justified by the accuracy of your data. On the other hand, if you have only 37 cases, and attempted to measure respondents' attitudes to work, then using two decimal places would be completely misleading. When presenting figures, less is usually more (helpful). And this is so, despite the default settings for output from analytical software. Take back control.

**Image 21.1**  Interminable numbers

It would also be helpful to offer the reader a sense of the relative error in each figure that you present. This is not the same as a measure of probabilistic uncertainty, which we might also want but do not have except in very rare circumstances. Chapter 7 explains why confidence intervals are so widely misunderstood (and therefore misleading), and that they are not measures of uncertainty in the result they are associated with. If anything, they are a kind of trustworthiness estimate, based on variation and scale, and there are much simpler versions of these that should be used instead (Gorard, 2019b). So the best indicator of doubt could be a sensitivity figure (see Chapter 12), or a range around the figure based on the estimated maximum relative error. The latter is what most audiences prefer, as in 'the unemployment rate rose to an estimated 3.9% (i.e. between 3.7% and 4.1%)' (van der Bles et al., 2020).

## PRESENTATION OF TABLES AND GRAPHS

Tables are the staple method for presenting multiple numeric results. Tables of simple figures can be extremely useful and informative. They present more precise figures than graphs, and can be designed so that the message for the reader is easy to spot. This is the key overarching point for both tables and graphs. Their purpose is to be part of a larger narrative. They should illustrate an important point in that narrative, where that point cannot be made as easily via text. This narrative point could be that value A is growing over time, or lower than value B, or the best predictor of value C, to list just a few examples.

As far as possible, there should be only one figure per table cell, nothing in brackets within any cell, and ideally only a maximum of two dimensions, represented in a two-dimensional table. The number of cases must be obvious in or near the table, or in the accompanying text. The row and column headers should be clear and meaningful. You might have to use acronyms or abbreviated variable names in your dataset, or capital letter names to suit your software. But these should not automatically become row titles or column headers. Put another way, your results should not be undigested output from analytical software (whether R, Stata, SPSS or whatever).

Note that some of the imaginary examples in other chapters in this book have used analytical output or variable names as headers, because they were used to help explain what technical terms mean, and how to do the exercises. This is not how they should be portrayed if they were presenting substantive results. So, for example, in reporting I would not head a table column 'Exp($B$)' (see Chapter 19). Instead, I would refer to it as the 'odds' (of something).

Many of the same points apply to graphs and their labelling. Use the correct format graph for the point you want the reader to grasp (easily). Generally, do not use pie charts, 3D graphs of any sort, or too many colours or formats. The main menu for everyday choice should be histograms/bar charts, scatterplots or line graphs. Do not have many lines in one graphic display, or lots of categories within the towers of your histograms, and so on. People will not understand them.

## REAL-LIFE EXAMPLES

Some of these points are illustrated with two real-life examples from research reports.

### Comparing means over time

This example is from a study of student career choice (Kunnen, 2013), selected because it is in an open access journal, and presents relatively simple results. The paper is no better or worse than thousands of others across the social sciences. It is just an example. The main results of the study were portrayed as follows:

> The commitment scores of the program participants increased significantly in the vocational and personal domain and in global identity [Table 21.2]. We calculated Cohen's *d* as an indication of effect size for the commitment scores that showed a significant increase, and in all cases a median effect size was found. The levels of exploration decreased significantly in the vocational domain and in the global identity.

**Table 21.2** Difference in the commitment and exploration scores per domain before (T1) and after (T2) the guidance (n = 45)

| Domain | T1 | St. dev. | T2 | St. dev. | T | sig. (1 tailed) | Cohen's d |
|---|---|---|---|---|---|---|---|
| **Commitment** | | | | | | | |
| Vocational | 22.8 | 6.72 | 26.7 | 6.31 | −3.573 | 0.003** | 0.60 |
| Personal | 22.8 | 6.66 | 25.7 | 7.56 | −2.590 | 0.007** | 0.41 |
| Global identity | 24.8 | 9.36 | 29.0 | 6.57 | −3.819 | 0.000*** | 0.52 |
| **Exploration** | | | | | | | |
| Vocational | 17.1 | 4.92 | 15.3 | 4.36 | 2.380 | 0.01** | |
| Personal | 13.6 | 5.32 | 13.6 | 4.78 | 0.096 | 0.92 | |
| Global identity | 12.3 | 6.00 | 11.1 | 4.91 | 2.710 | 0.01** | |

*P < 0.05; **P < 0.01; *** P < 0.001.
Source: Kunnen (2013: 5)

This is not a particularly complex table. There are no cells with two or more figures, for example, and the rows have real word descriptions, which are helpful. However, it is not easy for the reader to see the results as described above by the authors in this table. The column descriptions are mostly unhelpful. More importantly, the original description of the participants in the study was as follows:

> All 120 participants of the Career Guidance Program (Saxion Orientatie project) in 2009, 2010, and 2011 were invited to participate in the study. About half of them agreed and participated in the preproject interviews. Of these 60 participants, 15 did not participate in the second interview session after the program.

This means that there are only 45 cases for the main findings of changes over time (or 120/2 − 15). It is a good idea to put the number of cases in any analysis at the foot of the table, or in each row if this number varies within any set of results. The column labelled T has three decimal places, meaning that it is represented as being accurate to five ten-thousandths of a unit. This is impossible with only 45 cases, and the kind of measurement error that will occur when measuring a psychological construct such as a 'global identity'. Reducing the number of decimal places, perhaps to one as in the column labelled T1, would be more realistic and would make the table easier to read.

The study is an attempted population census (Chapter 10), which is seriously incomplete. There was no random sampling and no randomisation of cases to groups. Therefore, the use of significance tests and p-values is incorrect (even if it were useful). This is a very common but serious error in analysis (Chapter 6). Many studies you read that use numbers will probably make the mistake of presenting significance tests for cases that were never randomly selected, or for samples that have missing cases so that they are no longer random.

We should therefore ignore the gobbledegook at the foot of the table, and the columns labelled *T* (for the *t*-test as described in Chapter 6), and 'sig. (1 tailed)'. None of it makes sense in this non-randomised context, and can be safely ignored (and we can also ignore any reference to significance testing in the text of the paper).

It would make sense to compute the effect sizes for all six results, and not just the first three. The original study says it 'computed the effect size by means of Cohen's *d* (the difference between both measurements divided by the average standard deviation of both measurements)'. But Table 21.1 does not display either the 'difference between both measurements' or the average standard deviation of both measurements! This serious omission means that the reader cannot check the effect size calculation. This information would have been much more useful than knowing the value of *T* to three decimal places for a significance test that should not have been conducted. This information would be more important than almost anything else that *is* in the existing table. But it is missing.

However, because this is a before-and-after study, and we must assume that the results only apply to the 45 cases for whom there is before-and-after data (something that the paper should clarify), we can simply divide the total of the before and after standard deviations by 2 to get the overall standard deviation. Addressing all of these points leads to something like Table 21.3, accurate to one decimal place, with a new column (gain scores) showing the difference between before (originally labelled *T1*) and after (originally *T2*) scores, and another column showing the computed overall standard deviations.

**Table 21.3** Difference in the commitment and exploration scores per domain before and after the guidance, simpler

| Domain | Before scores | After scores | Gain scores | Average standard deviation | Cohen's d |
|---|---|---|---|---|---|
| Commitment | | | | | |
| Vocational | 22.8 | 26.7 | +3.9 | 6.5 | +0.6 |
| Personal | 22.8 | 25.7 | +2.9 | 7.2 | +0.4 |
| Global identity | 24.8 | 29.0 | +4.2 | 8.0 | +0.5 |
| Exploration | | | | | |
| Vocational | 17.1 | 15.3 | −1.8 | 4.7 | −0.40 |
| Personal | 13.6 | 13.6 | 0 | 5.1 | 0 |
| Global identity | 12.3 | 11.1 | −1.2 | 5.5 | −0.2 |

*n* = 45

If the purpose of the table is to help the reader understand the story of the findings, and because no comparison is made between the before-and-after scores for the different domains, the table could be even simpler (Table 21.4). In order to check the effect sizes we only need the difference between the before-and-after scores for each domain (the gain scores), and the average standard deviations of the before-and-after scores. The Cohen's *d* computation conducted by the authors was the gain score divided by the average standard deviation (see fuller explanation in Chapter 5). So for the vocational domain, 3.9/6.5 is 0.6, and so on. Of course, it may be important for some research reports to retain the before-and-after scores for some audiences (or present them first in a separate table).

It is also sometimes confusing to have different types of rows in a table (other than the header row), so the sub-headers ('Commitment' and 'Exploration') can be subsumed into the other six rows. Now we can see the pattern more clearly. Table 21.4 is much briefer than Table 21.2, but is actually more informative, makes the results easier to see, and stops the reader being misled by inappropriate significance testing. Using a simpler wording to express the results, it is now much easier to see the main results, which remain true without reference to the erroneous significance testing. The results are paraphrased here as follows:

> The reported commitment scores of the program participants increased noticeably in the vocational and personal domains and global identity. However, the reported levels of exploration either decreased or remained the same.

**Table 21.4** Difference in the commitment and exploration scores per domain before and after the guidance, even simpler

| Domain | Gain scores | Average standard deviation | Cohen's d |
| --- | --- | --- | --- |
| Vocational commitment | +3.9 | 6.5 | +0.6 |
| Personal commitment | +2.9 | 7.2 | +0.4 |
| Global identity commitment | +4.2 | 8.0 | +0.5 |
| Vocational exploration | −1.8 | 4.7 | −0.4 |
| Personal exploration | 0 | 5.1 | 0 |
| Global identity exploration | −1.2 | 5.5 | −0.2 |

Whether these results mean anything (or are useful) is a distinct question, and this is where the true analysis of data starts (Chapter 8). The results are now phrased in terms of 'reported' attitudes to emphasise to the reader that no actual behaviour was observed in this study. Using this phrasing would have been more accurate in the original. There are many other problems in this study, relating to the design and the conclusions drawn. Instead of spending time conducting and reporting meaningless, distracting significance tests, the author could have dealt with, or at least reported, these other problems and focused more on their own narrative and on making purported aids to understanding, such as tables, genuinely helpful.

Do you now see more of what was meant at the start of the book, about how dealing with numeric data may be hard work conceptually, and that you have to care about it, but that it is not complex or technical?

## Regression analysis

A slightly more complicated example of an unnecessarily confusing style of presentation comes from a regression analysis, used as part of a study by Çapri et al. (2013). Again it was chosen only because it is in an open access journal. The paper is no worse than thousands of others across the social sciences. It is just an example.

The study used 461 school students' reports of life satisfaction as the outcome variable, and three self-reported measures as predictors of life satisfaction. These predictors were attitude measures listed as hopelessness, absorption and efficacy. The results are presented as Table 21.5 (Table 4 in the original report). This is a somewhat difficult table to read. There are lots of cells, many of them empty but some with more than one figure, and some technical terms and abbreviations. Figures are presented to three decimal places.

**Table 21.5** Results of multiple regression analysis related to prediction of life satisfaction

| Predicted Variable | Analysis Phase | Predicting Variables | B | Standard Error | β | t | p | Zero-Order | $\Delta R^2$ |
|---|---|---|---|---|---|---|---|---|---|
| Life Satisfaction | | CONSTANT | 17.143 | 2.155 | | 7.954 | .000 | | |
| | | HOPELESSNESS | | | | | | | |
| | 1 | R= 0.416<br>$R^2$= 0.173 | −.515 | .067 | −.329 | −7.674 | .000 | −.416 | .173 |
| | | $F_{(1, 459)}$= 95.855* | | | | | | | |
| | | ABSORPTION | | | | | | | |
| | 2 | R= 0.489<br>$R^2$= 0.239 | .405 | .073 | .243 | 5.553 | .000 | .358 | .067 |
| | | $F_{(2, 458)}$= 72.076* | | | | | | | |
| | | EFFICACY | | | | | | | |
| | 3 | R= 0.496<br>$R^2$= 0.246 | −.136 | .067 | −.089 | −2.039 | .042 | −.254 | .007 |
| | | $F_{(3, 457)}$= 49.768* | | | | | | | |

**$p < .05$

Source: Çapri et al. (2013: 40)

The paper reports the participants thus:

> The research group is comprised of a total of 461 students ... who have voluntarily accepted to participate in the study and who continue their 12th grade education during the 2011-2012 school term at varying types of high schools.

There is no mention of how many of the cases approached refused to participate, and how many dropped out or provided unusable data (Chapter 8). The sample is clearly a convenience one – rather than randomised. This means that, as in the first example, there should be no reports of significance tests, standard errors or $p$-values. The columns labelled 'standard error', 't', and 'p' (the $p$-value) can therefore be deleted and safely ignored. They mean nothing in this context, and should not appear (see Chapter 6). We can also remove the footnote about $p$, and the rows containing $F$-values. Again these are all to do with significance tests, which are meaningless and misleading here.

The first column naming the outcome variable is covered by the table title, and the variable names for each 'analysis phase' can appear in the same line as their results. Given that there are only 461 cases, no estimate of missing data, and the variables are attitudinal scores, it also makes sense to reduce the number of decimal places to avoid suggesting a spurious accuracy. It is often good form to put a zero ahead of the decimal place when a number is only a fraction, and use plus and minus signs if figures can be either positive or negative. The constant can be noted as a footnote to the table, instead of having a nearly empty row just for that. And the column headers can be made more meaningful. Making all of these changes leads to the much simpler Table 21.6.

**Table 21.6** Results of multiple regression analysis related to prediction of life satisfaction, simpler

| Blocks | Variable entered in each phase | Regression coefficient B | Standardised coefficient beta | R-squared effect size | Increase in R-squared |
|---|---|---|---|---|---|
| 1 | Hopelessness | −0.52 | −0.33 | 0.17 | 0.17 |
| 2 | Absorption, Hopelessness | +0.41 | +0.24 | 0.24 | 0.07 |
| 3 | Efficacy. Absorption, Hopelessness | −0.14 | −0.09 | 0.25 | 0.01 |

Table 21.6 contains the same amount of useful information as Table 21.5. It is important to remember this. Nothing has been lost. Simplification is not the same as reducing useful information; it is only about presenting information better. The original table, with its disconcerting capital letters, looks like something lifted directly from analytical software output.

Here it is clearer to see that this is a regression model with three blocks (Chapter 17). The first block uses only one predictor (hopelessness), which can predict around 17% of the variation in life satisfaction. The second version uses two predictors, and the addition of absorption increases the predicted variation to 24%. Adding a third variable (efficacy) then makes little difference. Whether these three attitude variables can really be said to predict a fourth (life satisfaction), based on a survey in which all four attitudes were collected at the same time, is debatable. It is not clear what this association really means, or what use the results are. As before, looking for the meaning of the finding is the true analysis, and is what the report authors could have done with their time, instead of using significance tests.

## CONCLUSION

It matters that analysis is performed correctly, and running significance tests with non-random samples is just incorrect. Deleting all mention of this error from reports has many further advantages. It makes the description of methods and findings shorter, creating space for things that are too often omitted in reports (including some discussion of why the results matter – if they do). It reduces the task of the author, allowing them to focus instead on their narrative and on making purported aids to understanding, such as tables, genuinely helpful. This makes it easier to read any paper, and brings the writing into the purview of readers who do not cope well with inferential statistics (presumably with good reason). It is less likely to bamboozle readers with fake technicalities. And it should make it easier for the analyst to see what the true headline results are. There is no reason not to portray research as simply and accurately as possible. There is generally nothing to do *instead* of significance tests and the like. Their removal is usually improvement enough.

I hope that this book has helped you to understand more about the proper and sensible use of numbers in social science research, and to be more confident both when reading the studies of others and when doing your own research. This is important, because where statistical techniques are involved in research there is often little general understanding of their strengths and limitations. Even the kinds of effect sizes that this book describes have problems of misuse and misinterpretation. They, like all other aspects of working with numbers, require care, judgement and a sceptical frame of mind.

Good luck with those numbers.

### Suggestions for further reading

Some advice on how to simplify research reports that use numbers:
Wright, D. (2003) Making friends with your data: Improving how statistics are conducted and reported, *British Journal of Educational Psychology*, 73, 123-136.

This is an easy-to-read and entertaining introduction to the presentation of numeric information:
Huff, D. (1991) *How to Lie with Statistics*. Harmondsworth: Penguin.

This is a more recent, and slightly harder, book that also looks at the presentation and misrepresentation of numbers:
Levithin, S. (2017) *A Field Guide to Lies and Statistics*. London: Penguin.

A website that has interactive diagrams representing common statistical relationships:
https://rpsychologist.com/viz/

Two well-received and popular books on handling data:
Wheelan, C. (2013) *Naked Statistics: Stripping the Dread from the Data*. New York: W. W. Norton.

Spiegelhalter, D. (2019) *The Art of Statistics: How to Learn from Data*. New York: Basic Books.

# GLOSSARY

**Absolute value**  The modulus, or scale, of a number without its plus or minus sign.

**Autocorrelation**  The extent to which cases within a cluster (such as a prison or hospital) have similar characteristics.

**Bar chart**  A graph showing categories on the $x$-axis, and the frequencies of those categories on the $y$-axis.

**Bayes' theorem**  A formula for converting one conditional probability into its inverse, using the unconditional probabilities of both. $p(A|B)$ becomes $p(B|A)$ based on $p(A)$ and $p(B)$.

**Bias**  The extent to which an estimated figure differs from the 'true' or population figure, or a study is misleading.

**Binary**  Relating to two things, such as a variable with two categorical values

**Blinding**  To protect against internal bias and subversion, a research participant will not be aware of their allocated role in a research study. Can be applied to the researcher/analyst as well.

**Case**  A unit of analysis, such as a member of a sample or population.

**Categorical**  A variable type that identifies categories (such as ethnic group). Even if the variable values are expressed numerically, they are not real numbers.

**Census**  A study of a complete population.

**Central tendency (measures of)**  A general term for averages such as the mean, median and mode.

**Cluster randomised sampling**  Forming a sample in which cases are selected randomly from within clusters also selected randomly from the population. Ambiguous usage, and for some commentators includes simple samples of clusters.

**Cluster sampling**  Forming a sample in which the cases are groups or institutions, but measures are taken from units within each group (such as patients within hospitals).

**Cohort study**   A longitudinal design, following a group of cases and collecting data from them on more than one occasion. Often based on people born in the same short period of time.

**Continuous variable**   A real number measure based on a continuum, and so a measure in which there is no theoretical smallest unit (unlike a count).

**Control group**   A group in an experiment which is not exposed to the intervention. Often 'business as usual'.

**Counterfactual**   An alternative scenario, and the evidence from it. For example, there is little point in knowing that families living in poverty fared better after the implementation of new policy, without also considering what would have happened if the new policy did not exist.

**Cronbach's alpha**   A measure of reliability commonly used to assess the extent to which a group of questions are asking for the same basic underlying information in slightly different ways.

**Cross-plot**   A scatterplot graph, comparing pairs of values from the same cases for two variables, one on each axis.

**Cross-sectional design**   A snapshot study taking place in one time period, often involving a comparison of two or more subgroups.

**Cut-off point**   (or cut point) A point of division or separation on a scale of continuous values.

**Degrees of freedom**   The number of scores that are free to vary in a table, before the other cells are determined by the need to match the row and column totals.

**Dependent variable**   The outcome variable being predicted by one or more or independent variables, often in a regression model.

**Discrete variable**   A measure in which the units of measurements go up in jumps (like the price of a car in any currency), rather than on a continuous scale (like a temperature reading).

**Dispersion (measures of)**   A general term for estimates of how spread out a set of scores is. Includes standard deviation, mean absolute deviation, and inter-quartile range.

**Distribution**   The range and frequencies of a set of numbers, or their probability of occurrence. A uniform distribution, for example, has a range of values that are all equally likely.

**Effect size**   A scaled measure of the difference or association between two groups. Most commonly the difference between two means divided by their standard deviation.

**Epistemology**  A branch of philosophy concerned with how we can distinguish between opinions and more justified claims to knowledge (a bit like how we distinguish between real and fake news).

**Error component**  The part of any measurement or model that is not valid, and could be bias.

**Error propagation**  The way in which the error component in any measurements can increase as arithmetic is done with the measurements.

**Expected value**  In a cross-tabulation, the amount expected in one cell assuming a perfectly even distribution of figures by row and column.

**Experimental design**  Evaluation of a manipulated intervention where at least one group receives the treatment, and at least one does not. Cases are usually randomised to receive the intervention or not.

**FSM**  Eligibility for a free school meal is an official UK measure of living in poverty. The USA uses free or reduced-price lunch, and other countries have similar systems.

**Generalisability**  See *generality*.

**Generality**  The extent to which a (valid) research result might also be true for, or relevant to, cases and contexts not involved in the study.

**Heterogeneous groups**  Subgroups used in an analysis that are not intended to be matched or similar, such as ethnic groups. See also *homogeneous groups*.

**Histogram**  A graph that groups values into ranges on the $x$-axis, and shows the frequency of each range on the $y$-axis.

**Homogeneous groups**  Subgroups used in an analysis that are intended to be similar or balanced, such as those used at the outset in an experiment.

**Imputation**  Replacing a missing value in a dataset with a substitute value, computed from the cases that are not missing values.

**Independent variable**  A predictor variable used to predict the outcome variable, often in a regression model.

**Intention to treat**  A form of analysis for pragmatic evaluations (such as field experiments) in which cases are analysed as part of the groups to which they were initially allocated, even though they may have dropped out or changed groups.

**Interaction**  Two or more variables acting in combination. For example, the importance of having qualifications might vary depending upon your area of residence.

**Inter-quartile range**  If a set of data is ranked in terms of a value, and then divided into four equal-sized groups, then one quarter (quartile) will be those cases with the lowest 25% of scores, and another quarter will have the highest 25% scores. The middle two quarters form the 'inter-quartile', and the difference between the highest and lowest scores in that middle 50% scores form the range. It is a measure of dispersion like the standard deviation, but not as widely used.

**Interval measure**   A score or value based on a real number, so that each value on the scale of measurement is an equal interval from the next value.

**Isomorphism**   The situation where two concepts or processes can be mapped onto each other. For example, there should be an isomorphism between numbers used as measurements and the behaviour of the things they are measuring (such as going up or down in proportion). If this does not exist, the number is a pseudo-measurement.

**Latent variable**   Literally a hidden variable. The term is used when the surface data collected is used to create/envisage an underlying value, such as in factor analysis.

**Longitudinal design**   A study involving a sequence of data collection episodes taking place with the same cases over time.

**Mean**   A form of average. The sum of a set of measurements divided by the number of measurements.

**Mean absolute deviation**   A measure of the spread of values around the mean. It is the mean of all absolute deviations from the mean.

**Median**   A form of average. If the measurements are placed in size order, the median is the middle value.

**Mode**   A form of average. It is the most frequently occurring value in the dataset.

**Model**   A method of constructing a statistical link between variables to assess their interrelationship.

**Modulus**   See *absolute value*.

**Natural experiment**   Evaluation of an intervention that is not manipulated by the researcher (such as a new government policy). This usually means that cases cannot be randomly allocated to control and treatment groups, so a weaker design has to be used. An example of a quasi-experiment.

**Nominal measure**   A score or value from a scale which is not based on real numbers and in which the order is a matter of convention only.

**Non-probability sample**   A set of cases selected to take part in research without any kind of random selection. Could be an opportunity or convenience sample.

**Normal distribution**  A bell-shaped symmetrical frequency distribution that may underlie many social and psychological phenomena (such as height, income or test scores).

**Null hypothesis**  A hypothesis to be 'tested' or evaluated by examining the data achieved in a study compared to a set of expected results. This is usually taken to refer to the hypothesis of no observed difference, pattern or trend in the 'expected' results.

**Observed value**  In a cross-tabulation, the amount actually observed in the data for one cell. It is compared to the expected value.

**Odds ratio**  An effect size for use with 2 × 2 tables of categorical values. It involves finding the value for the product (multiplication) of the first and last cells, divided by the product of the other two cells.

**One-tailed prediction**  In comparing two groups, making clear before collecting data which set of scores is predicted to be the larger.

**Ordinal measure**  A score or value from a scale in which an order is clear, but which is not based on real numbers. A common example would be a Likert-type scale used in assessing attitudes.

**Outcome**  A dependent variable which can be assessed by manipulating independent or associated variables.

**Outlier**  A score in your results clearly outside the range of normal frequencies (surprisingly large or small).

**Paradigm**  In social science, this originally referred to a set of agreed rules that converted a research problem into an empirical puzzle that can be solved within those rules. 'What life is' might be a problem to solve, whereas the human genome project would be a puzzle, within a more restricted paradigm. Paradigm is now often (mis)used to refer to supposed approaches like qualitative or quantitative research, based on world views!

**Parsimony**  As used with regard to explanations, this refers to the explanation making the fewest additional assumptions. An explanation that requires us to imagine that something else is true even though we do not have clear evidence for it would be less parsimonious than one that was based only on widely accepted premises. Unparsimonious is therefore not the same as invalid or incorrect.

**Placebo**  A false treatment given to subgroups in an experimental trial, to help preserve a blind situation.

**Population**  All possible cases of interest to a study, from which a sample may be selected.

**Predictor**   A variable that explains some of the variation in an outcome or dependent variable.

**Probability**   The likelihood or chance of an occurrence, or of a number of possible outcomes.

**Proxy**   A variable that acts instead of another one in a regression model or explanation, predicting the variation that is perhaps more properly explained by the second variable. For example, an economic inequality might appear as a racial or neighbourhood inequality, or vice versa.

**Qualitative research**   Usually a term that refers to any research not involving numbers (such as text or sensory data). Supposedly based on a different world view/paradigm than quantitative research, but see Chapter 1.

**Quantitative research**   Usually a term that refers to any research involving numbers. Supposedly based on a different world view/paradigm than qualitative research, but see Chapter 1.

**Quasi-experimental design**   Like an experimental, or randomised control trial, design but without randomisation of cases to groups. Groups could be matched on known characteristics or naturally occurring. This makes the design weaker and the result less trustworthy (all other things being equal) than an experiment.

**Quota sample**   Like a stratified sample but without necessarily having a defined population. The researcher decides in advance that a certain proportion of the cases must have specific characteristics (such as age or sex).

**Random**   This describes an event occurring purely by chance.

**Random sample**   A set of cases selected for a research study by chance (randomly).

**Randomised control trial**   (RCT, sometimes 'controlled') Evaluation of an intervention which is manipulated so that at least one randomly allocated subgroup of cases receives the treatment and at least one does not.

**Ratio measure**   An example of an interval measure, with a value based on a real number and equal intervals on the scale of measurement, where the scale includes a real value of zero. Like a metre rule.

**Real number**   A variable type based on counts or measurements. It makes sense to do arithmetic with such values.

**Reliability**   An assessment of the extent to which a question, instrument or measure gives the same substantive result on different occasions or when taken by different means or people.

**Residual**  In regression analyses, a predicted score for the dependent variable is based on the value of the independent variable(s). Because of measurement and sampling error, and omission of key predictor variables, there will be a discrepancy between the observed and predicted score for each case. This is called the residual.

**Rounding**  Reducing the number of digits in a number, by making the number equal to the nearest of the least significant remaining digits. For example, 0.057 would be rounded to 0.06 using two decimal places.

**Sample**  The cases involved in a study, deliberately selected to represent a wider population of interest.

**Significant figures**  In a decimal fraction, for example, these are the digits after and including the first non-zero digit. The fraction 0.00038 has two significant figures.

**Standard deviation**  The square root of the variance. It is a slightly distorted summary of the average difference between each score in a set of results and the overall mean.

**Standard error**  The standard deviation of the distribution of repeated sample statistics. Theoretically, the standard error of the mean is a summary of the average difference between any specific sample mean and the overall mean of all means taken from equivalent samples. This can never be known in practice with only one actual sample to work with.

**Standardised coefficient**  The version of the coefficient used in a multivariate analysis, which assumes all of the predictor variables have been converted to standard z-scores.

**Stratified random sampling**  Sampling where the population is divided into subgroups (strata), each of which is sampled separately.

**Syntax**  A sequence of steps, like computer code, to produce an analysis.

**Treatment group**  A group in an experiment, which is exposed to the intervention (or treatment).

**Two-tailed prediction**  In comparing two groups, not making clear before collecting data which set of scores is predicted to be the larger.

**Validity**  The strength or trustworthiness of a measurement, study or result.

**Variance**  The sum of the squared deviations from the mean value; that is, the square of the standard deviation.

**Warrant**  A logical argument leading from research findings to the conclusions drawn by the researcher.

**Weight**  A value used to try and correct for a deficiency in the sample, by boosting the scores of one or more small subgroups.

**x-axis**  The horizontal axis on a graph, often portraying what is thought of as the independent variable.

**y-axis**  The vertical axis on a graph, often portraying what is thought of as the dependent variable.

**z-scores**  For any data item, its distance from the mean score, measured as a multiple of numbers of standard deviations. This creates a common 'currency' (of units of standard deviations) for all values in a model.

# REFERENCES

Abel, E., Sokol, R., Kruger, M. and Yargeau, D. (2008) Birthdates of medical school applicants, *Educational Studies*, 34(4), 271–275.

Aitkin, M. and Longford, N. (1986) Statistical modelling in school effectiveness studies, *Journal of the Royal Statistical Society Series A*, 149(3), 1–43.

Allison, P. (1984) *Event History Analysis: Regression for Longitudinal Event Data*. London: Sage.

Amir, E. (2012) On uses of mean absolute deviation: Decomposition, skewness and correlation coefficients, *METRON – International Journal of Statistics*, LXX(2–3), 145–164.

Anand, M. and Narasimha, Y. (2013) Removal of salt and pepper noise from highly corrupted images using mean deviation statistical parameter, *International Journal on Computer Science and Engineering*, 5(2), 113–119.

Ballatore, R., Paccagnella, M. and Tonello, M. (2016) *Bullied because younger than my mates? The effect of age rank on victimization at school*. www.bancaditalia.it/pubblicazioni/altri-atti-convegni/2016-human-capital/Ballatore_Paccagnella_Tonello.pdf

Barnett, V. and Lewis, T. (1978) *Outliers in Statistical Data*. Chichester: John Wiley & Sons.

Behaghel, L., Crepon, B., Gurgand, M. and Le Barbanchon, T. (2009) *Sample Attrition Bias in Randomized Surveys: A Tale of Two Surveys*, IZA Discussion Paper 4162. Bonn: Institute for the Study of Labor. http://ftp.iza.org/dp4162.pdf

Berchtold, A. (2019) Treatment and reporting of item-level missing data in social science research, *International Journal of Social Research Methodology*, 22(5), 431–439.

Berk, R. (2010) What you can and can't properly do with regression, *Journal of Quantitative Criminology*, 26(4), 481–487.

Berk, R. and Freedman, D. (2001) *Statistical assumptions as empirical commitments*. www.stat.berkeley.edu/~census/berk2.pdf

Berka, K. (1983) *Measurement: Its Concepts, Theories and Problems*. New York: Springer.

Berry, W. and Feldman, S. (1985) *Multiple Regression in Practice*. London: Sage.

Blalock, H. (1964) *Causal Inferences in Nonexperimental Research*. Chapel Hill: University of North Carolina Press.

Bland, J. and Altman, D. (1997) Statistics notes: Cronbach's alpha, *British Medical Journal*, 314, 275.

Bland, M. (2003) *Cluster randomised trials in the medical literature*. http://epi.klinikum.uni-muenster.de/StatMethMed/2003/Freiburg/Folien/MartinBland.pdf

Brighton, M. (2000) Making our measurements count, *Evaluation and Research in Education*, 14(3 & 4), 124–135.

Brignell, J. (2000) *Sorry, wrong number! The abuse of measurement*. European Science and Environment Forum.

Brunton-Smith, I., Carpenter, J., Kenward, M. and Tarling, R. (2014) Multiple imputation for handling missing data in social research, *Social Research Update*, 65(Autumn). https://sru.soc.surrey.ac.uk/SRU65.pdf

Cahan, S. and Gamliel, E. (2011) First among others? Cohen's d vs. alternative standardized mean group difference measures, *Practical Assessment, Research and Evaluation*, 16(10), 1–6.

Çapri, B., Gündüz, B. and Akbay, S. (2013) The study of relations between life satisfaction, burnout, work engagement and hopelessness of high school students, *International Education Studies*, 6(11), 35–46. www.ccsenet.org/journal/index.php/ies/article/view/29023/18264

Carver, R. (1978) The case against statistical significance testing, *Harvard Educational Review*, 48, 378–399.

Child, D. (1970) *The Essentials of Factor Analysis*. London: Holt, Rinehart and Winston.

Clegg, F. (1992) *Simple Statistics: A Course Book for the Social Sciences*. Cambridge: Cambridge University Press.

Cochrane (2012) http://www.cochrane-net.org/openlearning/html/modA2-4.htm

Cohen, J. (1977) *Statistical Power Analysis for the Behavioral Sciences*. London: Routledge.

Colquhoun, D. (2014) An investigation of the false discovery rate and the misinterpretation of p-values, *Royal Society Open Science*, 1, 1–16.

Comrey, A. (1973) *A First Course on Factor Analysis*. London: Academic Press.

Consort (n.d.) The CONSORT diagram. http://www.consort-statement.org/consort-statement/flow-diagram

Cook, T. and Campbell, D. (1979) *Quasi-experimentation: Design and Analysis Issues for Field Settings*. Chicago: Rand McNally.

Cox, D. (2001) Another comment on the role of statistical methods, *British Medical Journal*, 322, 231.

Crameri, A., von Wyl, A., Koemeda, M., Schulthess, P. and Tschuschke, V. (2015) Sensitivity analysis in multiple imputation in effectiveness studies of psychotherapy, *Frontiers in Psychology*, 6, 1042.

Crawford, C., Dearden, L. and Greaves, E. (2011) *Does when you are born matter? The impact of month of birth on children's cognitive and non-cognitive skills in England*, Report to the Nuffield Foundation.

Cronbach L. (1951) Coefficient alpha and the internal structure of tests, *Psychomerika*, 16, 297–334.

Cuddeback, G.. Wilson, E., Orme, J. and Combs-Orme, T. (2004) Detecting and statistically correcting sample selection bias, *Journal of Social Service Research*, 30(3), 19–30.

Cumming, G. (2013) The new statistics: Why and how, *Psychological Science*, 25(1), 7–29.

Cureton, E. and D'Agostino, R. (1983) *Factor Analysis: An Applied Approach*. Hillsdale, NJ: Lawrence Erlbaum.

Darroch, J. (1974) Multiplicative and additive interactions in contingency tables, *Biometrika*, 61(2), 207–214.

Dawes, R. (2001) *Everyday Irrationality*. Boulder, CO: Westview Press.

de Vaus, D. (2001) *Research Design in Social Research*. London: Sage.

de Vaus, D. (2002) *Analyzing Social Science Data: 50 Key Problems in Data Analysis*. London: Sage.

Department for Education (2016) New assessments and headline measures in 2016. Statistical First Release SFR 62/2016. https://www.gov.uk/government/uploads/system/uploads/attachment_data/file/577296/SFR62_2016_text.pdf

Dolton, P., Lindeboom, M. and Van den Berg, G. (2000) *Survey attrition: A taxonomy and the search for valid instruments to correct for biases*. http://www.fcsm.gov/99papers/berlin.html

Erikson, R. and Goldthorpe, J. (1991) *The Constant Flux: A Study of Class Mobility in Industrial Societies*. Oxford: Clarendon Press.

Everitt, B. (1980) *Cluster Analysis*. London: Heinemann.

Everitt, B. and Smith, A. (1979) Interactions in contingency tables: A brief discussion of alternative definitions, *Psychological Medicine*, 9, 581–583.

Falk, R. and Greenbaum, C. (1995) Significance tests die hard: The amazing persistence of a probabilistic misconception, *Theory and Psychology*, 5, 75–98.

Fidler, F., Thomson, N., Cumming, G., Finch, S. and Leeman, J. (2004) Editors can lead researchers to confidence intervals, but can't make them think: Statistical reform lessons from medicine, *Psychological Science*, 15(2), 119–126.

Field, A. (2013) *Discovering Statistics Using IBM SPSS Statistics*. London: Sage.

Fitz-Gibbon, C. (2001) *Value-Added for Those in Despair: Research Methods Matter*. Newbury Park, CA: Sage.

Frank, K., Maroulis, S., Doung, M. and Kelcey, B. (2013) What would it take to change an inference? Using Rubin's causal model to interpret the robustness of causal inferences, *Educational Evaluation and Policy Analysis*, 35(4), 437–460.

Giacquinta, J. and Shaw, F. (2000) *Judging non-returner induced sample bias from the study of early and late returners: A useful approach?* Presentation at AERA Conference, New Orleans, April.

Gigerenzer, G. (2002) *Reckoning with Risk*. London: Penguin.

Gilbert, N. (1981) *Modelling Society: An Introduction to Loglinear Analysis for Social Researchers*. London: George Allen & Unwin.

Gilbert, N. (1993) *Analysing Tabular Data: Loglinear and Logistic Models for Social Researchers*. London: UCL Press.

Glass, G. (2014) Random selection, random assignment and Sir Ronald Fisher, *Psychology of Education Review*, 38(1), 12–13.

Goldthorpe, J., Llewellyn, C. and Payne, C. (1987) *Social Mobility and Class Structure in Modern Britain*. Oxford: Clarendon Press.

Gorard, S. (1999) Keeping a sense of proportion: The 'politician's error' in analysing school outcomes, *British Journal of Educational Studies*, 47(3), 235–246.

Gorard, S. (2000) *Education and Social Justice*. Cardiff: University of Wales Press.

Gorard, S. (2003a) Understanding probabilities and re-considering traditional research methods training, *Sociological Research Online*, 8, 1.

Gorard, S. (2003b) *Quantitative Methods in Social Science: The Role of Numbers Made Easy*. London: Continuum.

Gorard, S. (2005) Revisiting a 90-year-old debate: The advantages of the mean deviation, *British Journal of Educational Studies*, 53(4), 417–430.

Gorard, S. (2006a) *Using Everyday Numbers Effectively in Research: Not a Book about Statistics*. London: Continuum.

Gorard, S. (2006b) Towards a judgement-based statistical analysis, *British Journal of Sociology of Education*, 27(1), 67–80.

Gorard, S. (2007) The dubious benefits of multi-level modelling, *International Journal of Research and Method in Education*, 30(2), 221–236.

Gorard, S. (2010a) Measuring is more than assigning numbers. In G. Walford, E. Tucker and M. Viswanathan (eds), *Sage Handbook of Measurement* (pp. 389–408). Thousand Oaks, CA: Sage.

Gorard, S. (2010b) All evidence is equal: The flaw in statistical reasoning, *Oxford Review of Education*, 36(1), 63–77.

Gorard, S. (2010c) Serious doubts about school effectiveness, *British Educational Research Journal*, 36(5), 735–766.

Gorard, S. (2012) Who is eligible for free school meals? Characterising FSM as a measure of disadvantage in England, *British Educational Research Journal*, 38(6), 1003–1017.

Gorard, S. (2013a) The propagation of errors in experimental data analysis: A comparison of pre- and post-test designs, *International Journal of Research and Method in Education*, 36(4), 372–385.

Gorard, S. (2013b) *Research Design: Robust Approaches for the Social Sciences*. London: Sage.

Gorard, S. (2015a) Introducing the mean absolute deviation 'effect' size, *International Journal of Research and Method in Education*, 38(2), 105–114.

Gorard, S. (2015b) The complex determinants of school intake characteristics, England 1989 to 2014, *Cambridge Journal of Education*, 46(1), 131–146.

Gorard, S. (2017) How prepared do newly-qualified teachers feel? Differences between routes and settings, *Journal of Education for Teaching*, 43(1), 3–19.

Gorard, S. (2018) *Education Policy: Evidence of Equity and Effectiveness*. Bristol: Policy Press.

Gorard, S. (2019a) Significance testing with incompletely randomised cases cannot possibly work, *International Journal of Science and Research Methodology*, 11(2), 42–51.

Gorard, S. (2019b) Do we really need confidence intervals in the new statistics? *International Journal of Social Research Methodology*, 22(3), 281–291.

Gorard, S. (2020) Handling missing data in numeric analyses, *International Journal of Social Research Methods*, 23(6), 651–660. https://www.tandfonline.com/doi/full/10.1080/13645579.2020.1729974

Gorard, S. and Rees, G. (2002) *Creating a Learning Society*. Bristol: Policy Press.

Gorard, S. and Siddiqui, N. (2019) How trajectories of disadvantage help explain school attainment, *SAGE Open*. https://journals.sagepub.com/doi/10.1177/2158244018825171

Gorard, S. and Taylor, C. (2002) What is segregation? A comparison of measures in terms of strong and weak compositional invariance, *Sociology*, 36(4), 875–895.

Gorard, S. with Taylor, C. (2004) *Combining Methods in Educational and Social Research*. Maidenhead: Open University Press.

Gorard, S., Rees, G. and Salisbury, J. (2001) The differential attainment of boys and girls at school: Investigating the patterns and their determinants, *British Educational Research Journal*, 27(2), 125–139.

Gorard, S., Taylor, C. and Fitz, J. (2003) *Schools, Markets and Choice Policies*. London: RoutledgeFalmer.

Gorard, S., Siddiqui, N. and See, B. H. (2015) An evaluation of the 'Switch-on reading' literacy catch-up programme, *British Educational Research Journal*, 41(4), 596–612.

Gorard, S., See, B. H. and Siddiqui, N. (2017) *The Trials of Evidence-Based Education*. London: Routledge.

Gorard, S., Siddiqui, N. and See, B. H. (2019) The difficulties of judging what difference the Pupil Premium has made to school intakes and outcomes in England, *Research Papers in Education*. https://doi.org/10.1080/02671522.2019.1677759

Gorard, S., Ventista, O., Morris, R. and See, B. H. (2020) *Who wants to be a teacher? Findings from a survey of undergraduates in England*, Durham University Evidence Centre for Education. https://www.researchgate.net/publication/339537756_Who_wants_to_be_a_teacher_Findings_from_a_survey_of_undergraduates_in_England?channel=doi&linkId=5e57ed3f92851cefa1c9ce18&showFulltext=true

Gorsuch, R. (1972) *Factor Analysis*. London: Saunders.

Gough, D. (2007) Weight of evidence: A framework for the appraisal of the quality and relevance of evidence, *Research Papers in Education*, 22(2), 213–228.

Hagenaars, J. (1990) *Categorical Longitudinal Data: Log-linear, Panel, Trend and Cohort Analysis*. London: Sage.

Hansen, M. and Hurwitz, W. (1946) The problem of non-response in sample surveys, *Journal of the American Statistical Association*, 41, 517–529.

Hao, Y., Flowers, H., Monti, M. and Qualters, J. (2012) U.S. census unit population exposures to ambient air pollutants, *International Journal of Health Geographics*, 11, 3.

Harlow, L., Mulaik, S. and Steiger, J. (1997) *What If There Were No Significance Tests?* Mahwah, NJ: Lawrence Erlbaum.

Harwell, M. and Gatti, G. (2001) Rescaling ordinal data to interval data in educational research, *Review of Educational Research*, 71(1), 105–131.

Hedges, L. and Olkin, I. (1985) *Statistical Methods for Meta-analysis*. San Diego, CA: Academic Press.

Huber, P. (1981) *Robust Statistics*. New York: John Wiley & Sons.

Hughes, R., Heron, J., Sterne, J. and Tilling, K. (2019) Accounting for missing data in statistical analyses: Multiple imputation is not always the answer, *International Journal of Epidemiology*, 48(4), 1294–1304.

Ioannidis, J. (2005) Why most published research findings are false, *PLoS Medicine*, 2(8), e124. http://www.ncbi.nlm.nih.gov/pmc/articles/PMC1182327/

Ioannidis, J. (2019) What have we (not) learnt from millions of scientific papers with p values? *The American Statistician*, 73(sup1), 20–25.

Jackson, D. and Borgatta, E. (1981) *Factor Analysis and Measurement*. London: Sage.

Kashimada, K. and Koopman, P. (2010) Sry: The master switch in mammalian sex determination, *Development*, 137(23), 3921–3930.

Kolmogorov, A. and Uspenskii, V. (1987) Algorithms and randomness, *Theory of Probability and its Applications*, 32(3), 389–412.

Kunnen, E. (2013) The effects of career choice guidance on identity development, *Education Research International*, 2013, 901718. http://dx.doi.org/10.1155/2013/901718

Lee, E., Forthofer, R. and Lorimor, R. (1989) *Analyzing Complex Survey Data*. London: Sage.

Lehtonen, R. and Pahkinen, E. (1995) *Practical Methods for Design and Analysis of Complex Surveys*. Chichester: John Wiley & Sons.

Lenhard, W. and Lenhard, A. (2016) *Calculation of effect sizes*. Dettelbach, Germany: Psychometrika. https://www.psychometrica.de/effect_size.html

Lieberson, S. (1981) An asymmetrical approach to segregation. In C. Peach, V. Robinson and S. Smith (eds), *Ethnic Segregation in Cities*. London: Croom Helm.

Lindner, J., Murphy, T. and Briers, G. (2001) Handling non-response in social science research, *Journal of Agricultural Education*, 42(3), 43–53.

Lipsey, M., Puzio, K., Yun, C., Hebert, M., Steinka-Fry, K., Cole, M., Roberts, M., Anthony, K. and Busick, M. (2012) *Translating the Statistical Representation of the Effects of Education Interventions into More Readily Interpretable Forms*. Washington, DC: Institute of Education Sciences.

Little, R. and Rubin, D. (2002) *Statistical Analysis with Missing Data*, 2nd edition. Hoboken, NJ: John Wiley & Sons.

Madaleno, M. and Waights, S. (n.d.) *Guide to scoring methods using the Maryland Scientific Methods Scale.* https://whatworksgrowth.org/public/files/Scoring-Guide.pdf

Mare, R. and Winship, C. (1985) School enrollment, military enlistment, and the transition to work: Implications for the age pattern of employment. In J. Heckman and B. Singer (eds), *Longitudinal Analysis of Labor Market Data.* Cambridge: Cambridge University Press.

Marradi, A. (1981) Factor analysis as an aid in the formation and refinement of empirically useful concepts. In D. Jackson and E. Borgatta (eds), *Factor Analysis and Measurement.* London: Sage.

Marshall, G., Swift, A. and Roberts, S. (1997) *Against the Odds? Social Class and Social Justice in Industrial Societies.* Oxford: Clarendon Press.

Maruyama, G. (1998) *Basics of Structural Equation Modelling.* London: Sage.

Massey, D. and Denton, N. (1988) The dimensions of residential segregation, *Social Forces*, 67, 373–393.

Matthews, R. (1998) Flukes and flaws, *Prospect*, 20 November. http://www.prospectmagazine.co.uk/features/flukesandflaws

Matthews, R. (2002) The cold reality of probability theory, *The Sunday Telegraph*, 5 May, p. 31.

Maxwell, A. (1977) *Multivariate Analysis in Behavioural Research.* London: Chapman & Hall.

McPhetres, J. and Pennycook, G. (2020) Lay people are unimpressed by the effect sizes typically reported in psychological science, *PsyArXiv*. doi: 10.31234/osf.io/qu9hn

Meehl, P. (1967) Theory-testing in psychology and physics: A methodological paradox, *Philosophy of Science*, 34(2), 103–115.

Melkonian, M. and Areepattamannil, S. (2017) The effect of absolute age-position on academic performance: A study of secondary students in the United Arab Emirates, *Educational Studies*, 44(5), 551–563. http://dx.doi.org/10.1080/03055698.2017.1382330

Menard, S. (1995) *Applied Logistic Regression Analysis.* London: Sage.

Miles, J. and Shevlin, M. (2001) *Applying Regression and Correlation: A Guide for Students and Researchers.* London: Sage.

Moses, L. (2001) *A larger role for randomized experiments in educational policy research*, Presentation at AERA annual conference, Seattle, April.

Norušis, M. (2000) *SPSS 10.0 Guide to Data Analysis.* Upper Saddle River, NJ: Prentice Hall.

Nunnally, J. (1975) Psychometric theory 25 years ago and now, *Educational Researcher*, 4(7), 7–21.

Oakley, A. (1998) Science, gender, and women's liberation: An argument against postmodernism, *Women's Studies International Forum*, 21(2), 133–146.

Pannell, D. (1997) Sensitivity analysis of normative economic models: Theoretical framework and practical strategies, *Agricultural Economics*, 16, 139–152.

Pearl, J. and Mackenzie, D. (2018) *The Book of Why: The New Science of Cause and Effect*. New York: Basic Books.

Pedhazur, E. (1982) *Multiple Regression in Behavioural Research*. New York: Holt, Rinehart and Winston.

Peers, I. (1996) *Statistical Analysis for Education and Psychology Researchers*. London: Falmer Press.

Peress, M. (2010) Correcting for survey nonresponse using variable response propensity, *Journal of the American Statistical Association*, 105(492), 1418–1430.

Phillips, D. (2014) Research in the grad sciences, and in the very hard 'softer' sciences, *Educational Researcher*, 43(1), 9–11.

Puffer, S., Torgerson, D. and Watson, J. (2005) Cluster randomized controlled trials, *Journal of Evaluation in Clinical Practice*, 11(5), 479–483.

Ralston, K., MacInnes, J., Crow, G. and Gayle, V. (2016) *We need to talk about statistical anxiety*, National Centre for Research Methods Working Paper 4/16. http://eprints.ncrm.ac.uk/3987/1/anxiety_literature_WP4_16.pdf

Raudenbush, S. (2002) *New directions in the evaluation of Title I*, Presentation at AERA annual meeting, New Orleans, April.

Raudenbush, S. and Bryk, A. (1986) A hierarchical model for studying school effects, *Sociology of Education*, 59(1), 1–17.

Reinertsen, A. (2015) A wild multiple apparatus of knowing discussion: Waiting pentimento intra-observation, *Qualitative Inquiry*, 22(4), 263–273.

See, B. H., Gorard, S. and Siddiqui, N. (2015) Can teachers use research evidence in practice? A pilot study of the use of feedback to enhance learning, *Educational Research*, 58(1), 56–72.

See, B. H., Morris, R., Gorard, S. and Siddiqui, N. (2020) *Writing about values*, Evaluation Report, Education Endowment Foundation. https://educationendowmentfoundation.org.uk/projects-and-evaluation/projects/writing-about-values/

Shadish, W., Cook, T. and Campbell, D. (2002) *Experimental and Quasi-experimental Designs for Generalized Causal Inference*. Belmont, CA: Wadsworth.

Sheikh, K. and Mattingly, S. (1981) Investigating nonresponse bias in mail surveys, *Journal of Epidemiology and Community Health*, 35, 293–296.

Siddiqui, N., Gorard, S. and See, B. H. (2015) Accelerated Reader as a literacy catch-up intervention during the primary to secondary school transition phase, *Educational Review*, 68(2), 139–154.

Siddiqui, N., Boliver, V. and Gorard, S. (2019) Reliability of longitudinal social surveys of access to higher education: The case of Next Steps in England, *Social Inclusion*, 7(1), 80–89.

Siegel, S. (1956) *Nonparametric Statistics for the Behavioural Sciences*. Tokyo: McGraw-Hill.

Siegfried, T. (2015). P value ban: Small step for a journal, giant leap for science, *Science News*, 17 March. https://www.sciencenews.org/blog/context/p-value-ban-small-step-journal-giant-leap-science

Soley-Bori, M. (2013) *Dealing with missing data: Key assumptions and methods for applied analysis*, Technical Report No. 4, Boston University School of Public Health. https://www.bu.edu/sph/files/2014/05/Marina-tech-report.pdf

Starbuck, W. (2016) 60th Anniversary Essay: How journals could improve research practices in social science, *Administrative Science Quarterly*, 61(2), 165–183.

Sterne, J., White, I., Carlin, J., Spratt, M., Royston, P., Kenward, M., Wood, A. and Carpenter, J. (2009) Multiple imputation for missing data in epidemiological and clinical research: Potential and pitfalls, *British Medical Journal*, 338, b2393.

Stevens, J. (1992) *Applied Multivariate Statistics for the Social Sciences*. Hillsdale, NJ: Lawrence Erlbaum.

Stevens, S. (1946) On the theory of scales of measurement, *Science*, 103(2684), 677–680.

Swalin, A. (2018) *How to handle missing data*. https://towardsdatascience.com/how-to-handle-missing-data-8646b18db0d4

Tarran, B. (2019) Is this the end of 'statistical significance'?, *Significance*, 16(2), 4.

Thabane, L., Mbuagbaw, L., Zhang, S., Samaan, Z., Marcucci, M., Ye, C., Thabane, M. et al. (2013) A tutorial on sensitivity analyses in clinical trials, *BMC Medical Research Methodology*, 13, 92. www.biomedcentral.com/1471-2288/13/92

van Buuren, S. (2018) *Flexible Imputation of Missing Data*. Boca Raton, FL: CRC Press.

van der Bles, A., van der Linden, S., Freeman, A. and Spiegelhalter, D. (2020) The effects of communicating uncertainty on public trust in facts and numbers, *Proceedings of the National Academy of Sciences of the USA*, 117(14), 7672–7683.

Volcan, S. (2002) What is a random sequence? *Mathematical Association of America Monthly*, 109, 46–63. https://www.maa.org/sites/default/files/pdf/upload_library/22/Ford/Volchan46-63.pdf

von Mises, R. and Doob, J. (1941) Discussion of papers in probability theory, *Annals of Mathematical Statistics*, 12(2), 215–217.

Watts, D. (1991) Why is introductory statistics difficult to learn? *The American Statistician*, 45(4), 290–291.

Wright, D. (2003) Making friends with your data: Improving how statistics are conducted and reported, *British Journal of Educational Psychology*, 73, 123–136.

Xiao, Z., Kasim, A. and Higgins, S. (2016) Same difference? Understanding variation in the estimation of effect sizes from trial data, *International Journal of Educational Research*, 77, 1–14.

# INDEX

absolute mean deviation 26–29
absolute value 27–28
achievement gap 54–55
adjusted R squared 198, 202, 211
ANOVA, *see* significance testing
assumptions for an analysis, 66–67, 69–70, 77–88, 183, 188–190, 195, 220–225
attitudes 13–14, 96, 170–171, 175, 251–253, 262–268
attrition, *see* missing data
auto-correlation 122, 223
average, *see* mean, median, mode

backward entry of variables, *see* forward entry
bar chart 23–24
Bayes' Theorem 83
bias 20–21, 78–80, 137–139, 147, 155, 166–168, 170–176
binary logistic regression, *see* logistic regression
blind 102

case study design 106
cases 23, 106–107, 116–117, 124–128, 136–137
categorical values 9–10, 19–24, 36–49, 191, 220, 226–243
cause and effect 106, 189–190, 203–204
census 116, 119
central tendency, *see* mean, median, mode
chi-squared distribution 66, 73
chi-squared test 63–67, 79
classification plot 233
classification table 230–231
cleaning data 20–21
cluster samples 121–122, 134–135, 222–225
coding scheme 20
coefficients 198, 203, 211, 231–232
Cohen's d, *see* effect size
cohort studies 78, 117
column percentages 40
complete case analysis 154–156
complete values analysis 156

composition invariance 48
conditional probabilities 70–72, 78–83, 258–259
confidence intervals 78, 84–87
CONSORT 144–145
continuous measures 15
control group 108
convenience sample 67, 122–124
correlation 60, 181–192, 195
costs 98, 102
counterfactual 108
counts 9–11
Cronbach's alpha, *see* reliability
cross-product ratio, *see* odds ratio
cross-sectional design 106–107
cross-tabulation 37–44

data preparation 19–21
degrees of freedom 37, 41, 66, 69
dependent variable 52, 58–59, 164, 196, 206
design, *see* research design
discrete values 15
disparity ratio 48
dispersion (of scores), *see* inter-quartile range, mean absolute deviation, standard deviation
distributions, *see* uniform, normal, chi-squared, t-test
dropout, *see* missing data
dummy variables 207–208, 212, 220

effect size 56–59, 71, 84, 98, 102, 125, 151–152, 185
epistemology 3–4
error propagation 155–156, 171–173, 242
ethnic group, *see* ethnicity
ethnicity 10, 45–46, 49, 125, 206
everyday use of numbers 3–4, 14
Excel 6, 22–23, 133
expected values 37–38, 41, 65–66
experiment 58–60, 84, 101–102, 106, 108–109, 146–148, 155, 215

F test, *see* significance testing
factor analysis 245–251
Fisher's exact test, *see* chi-squared test
forward entry of variables 209–211, 215, 220–221, 234–235
free school meals 44–46, 242
frequencies 20–24, 37–44
FSM, *see* free school meals

generalisability, *see* generality
generality 99, 114–115, 117–119
geography 28, 37–40, 47, 53–54, 72
Glass's delta, *see* effect size
graph, *see* bar chart, histogram, line graph, scatterplot

health science 47–49, 127
Hedge's g, *see* effect size
heterogeneous groups 108
histogram 25–26, 29, 31–32, 53–54, 57
homogeneous group 108
hospitals 22–26

imputation 157–158, 162–165
independent samples 67, 69
independent variable 52, 58–59, 164, 196, 206
interaction effects 204–206
interval values 12, 14–15
isomorphism 12

judgement 3, 5, 13, 58, 91–102

Likert scales 13–14
line graph 190–191
linear regression, *see* regression analysis
loadings (in factor analysis) 249–251
logistic regression 226–243
logit regression, *see* logistic regression
longitudinal design 106–107, 111, 155

matching cases 109
mean 25–29, 52–54, 59, 84–86
measure of spread, *see* inter-quartile range, mean absolute deviation, standard deviation
measurement 9–16, 93–94, 170–176
measurement error 78, 93–94, 96, 98, 125, 138–139, 155–156, 170–176, 221, 247
median 25
missing at random 166–168
missing data 20, 22, 40–41, 44–46, 66, 77, 93–94, 107–108, 115, 119, 122, 133–134, 137–158, 162–169
mode 22–23, 25
modulus, *see* absolute value
multiple imputation, *see* imputation
multiple regression, *see* regression analysis

natural experiment, *see* experiment
NNTD 150–152
nominal values 9–10
non-probability sample, *see* convenience sample
normal distribution 32–33, 57, 72, 85
null hypothesis 37, 81
number needed to disturb (a finding), *see* NNTD

observed values 38–39, 41–42, 66
occupational class scale 13, 47–49, 147, 220
odds ratios 48, 54, 60
one-sided or one-tailed test 66, 69
opportunity sample, *see* convenience sample
ordinal values 9–10, 12–13
outliers 19–21
oversampling 152

paradigms 3–5, 124
parsimony, *see* simplicity
Pearson chi-square, *see* chi-squared test
Pearson's correlation coefficient, *see* correlation
percentages 22–23, 37–44
placebo 143
population 66, 78, 81–83, 85, 96, 106, 114–129
post-intervention test 59, 109, 148
power calculations 78, 83–84, 127–128
predictor, *see* independent variable
pre-intervention test 59, 109, 148
presentation of results, *see* reporting research
principal components analysis, *see* factor analysis
probability 30
psychology 13, 72, 87, 124, 251, 263
publication bias 75
p-values, *see* significance testing

qualitative, *see* paradigms
quantitative, *see* paradigms
quasi-experiment, *see* experiment

random numbers 30–32, 59, 66, 80, 109, 120–121, 131–135, 165–168, 184–185
random sample 66–67, 69, 77–78, 85, 107, 120–123, 133–135
randomisation, *see* random numbers
randomised control(led) trial, *see* experiment
ratio values 14–15
RCT, *see* experiment
real numbers 9–11, 14, 24–30, 51–61, 181–192, 226, 232
reference category (in logistic regression) 240
regression analysis 78, 119, 164, 194–200, 201–216, 220–225, 266–268
reliability 174–176, 251–253

replacing missing values, *see* imputation
reporting research 20–21, 42, 44, 92–93, 97, 107, 144–145, 257–268
research design 92–94, 105–112, 118
research question 92, 94, 105–108
risk ratio 48
row percentages 40

sample 67, 81–83, 85, 114–129, 137–138
sample size, *see* scale
sampling, *see* sample
scale of research (N) 71, 85, 93–94, 96, 107, 115, 118, 124–128, 154–155
scatterplot 182, 184, 186, 195–196
scree plot 247
secondary data 45-46, 118
segregation ratio 49
SEN 46, 191
sensitivity analysis 148–152
SES, *see* FSM, occupational class
significance testing 63–75, 77–88, 128, 139, 156, 168, 183, 259
significant figures 260
simplicity 99, 173, 206–207, 209–211, 214, 247, 250, 257–268
snowball sample 123–124
social class, *see* occupational class
sociology 87

Special educational need or disability, *see* SEN
SPSS 6, 23
standard deviation 25, 28–29, 52–59, 68, 84–85, 98, 157
standard error 68, 78, 85, 156, 168
standardised coefficient, *see* coefficient
stratified sampling 121
Switch-on Reading 58–60, 148, 206–211, 215
syntax 6

t distribution 72–73
theory 98
total 25, 27
treatment
trial, *see* experiment
t-test 63, 67–69, 79, 81
two-sided or two-tailed test 66, 69

uniform distribution 30–31
univariate analysis 19–33
unstandardised coefficient, *see* coefficient

validity 20, 22, 92–99, 109, 114–115, 119, 124, 176
variance, *see* standard deviation

warrant 99–101, 215
weighting data 153–154